INDIVIDUALISM AND THE UNITY OF SCIENCE

The Worldly Philosophy: Studies at the Intersection of Philosophy and Economics

General Editor: Philip Mirowski, University of Notre Dame

Testing, Rationality, and Progress: Essays on the Popperian Tradition in Economic Methodology
by D. Wade Hands
Edgeworth on Chance, Economic Hazard, and Statistics
edited by Philip Mirowski
Intellectual Trespassing as a Way of Life: Essays in Philosophy, Economics, and Mathematics
by David P. Ellerman
Individualism and the Unity of Science: Essays on Reduction, Explanation, and the Special Sciences
by Harold Kincaid

INDIVIDUALISM AND THE UNITY OF SCIENCE

Essays on Reduction, Explanation, and the Special Sciences

Harold Kincaid

ROWMAN & LITTLEFIELD PUBLISHERS, INC.
Lanham • Boulder • New York • Oxford

ROWMAN & LITTLEFIELD PUBLISHERS, INC.

Published in the United States of America
by Rowman & Littlefield Publishers, Inc.
4720 Boston Way, Lanham, Maryland 20706

12 Hid's Copse Road
Cummor Hill, Oxford OX2 9JJ, England

British Library Cataloguing in Publication Information Available

Library of Congress Cataloging-in-Publication Data

Kincaid, Harold, 1952—
Individualism and the unity of science : essays on reduction, explanation, and the special sciences / Harold Kincaid.
p. cm.—(Worldly philosophy)
Includes bibliographical references and index.
ISBN 0-8476-8662-0 (alk. paper).—ISBN 0-8476-8663-9 (pbk. : alk. paper)
1. Science—Philosophy. 2. Whole and parts (Philosophy). 3. Reductionism—Controversial literature. 4. Individualism—Controversial literature. I. Title. II. Series.
B67.K52 1997
111′.82—dc21 97-19279
CIP

ISBN 0-8476-8662-0 (cloth : alk. paper)
ISBN 0-8476-8663-9 (pbk. : alk. paper)

Printed in the United States of America

∞ ™ The paper used in this publication meets the minimum requirements of American National Standard for Information Sciences—Permanence of Paper for Printed Library Materials, ANSI Z39.48–1984.

Contents

Acknowledgments

I am very grateful for the help of a great many people in completing this book. Karl Matlin was an enormous help with the ideas of Chapter 4 as was Tim Day for Chapter 6. Scott Arnold, Alex Rosenberg, Paul Teller, Scott Gordon, Geoffrey Hellman, John Post, Terry Horgan, Lee McIntyre, George Graham, Bob McCauley, Alison Wylie, Merrilee Salmon, and Dan Little all gave me careful comments at one stage or another on this project. My biggest debt is to Phil Mirowski, for both his encouragement and his careful comments.

I would also like to acknowledge the following journals for allowing me to reprint parts of my previous articles: *Philosophy of Science* for use of "Reduction, Explanation, and Individualism" (1986), December, 53: 492–513 and "Molecular Biology and the Unity of Science" (1990), December, 57: 575–93; *Synthese* for the use of "Supervenience and Explanation" (1988), March, 77: 251–81 and "The Empirical Nature of the Individualism-Holism Dispute" (1993), November, 97: 229–47; *The Southern Journal of Philosophy* for use of "Supervenience Doesn't Entail Reducibility" (1987), Fall, 25: 342–56 and "Can Neoclassical Economics Be Defended on Grounds of Explanatory Power?" (1995 Supplement) 34: 155–79.

1

The Roots of Reductionism

Presumably there is but one universe, a universe that is made up of material particles and forces. Yet the universe we know is complex, and we often seem to understand it best by leaving physics behind for biology, psychology, or economics. This book concerns how that can be—why it might be that a physical universe is best understood in nonphysical terms or, even more generally, why complex entities are often best understood in their own terms rather than through their parts. I approach these large issues most directly via the individualism–holism dispute in the social sciences. Individualists claim that we can understand everything we want to know about the social world entirely in terms of the actions of individuals. I argue that this thesis is false. Yet the individualism–holism dispute is of a piece with much bigger questions about the unity of science and the nature of explanation. After all, individualism is just the general reductionist program applied to the social sciences. In the end my target is thus this larger agenda. Though individualism is my focus, I do not hesitate to draw general morals about explanation and scientific unity.

This introductory chapter sets up the issues, explains why they are important, sketches a picture of a nonreductive yet unified science, and diagnoses the misguided historical sources of reductionism's appeal. My concern is to outline the issues and arguments to come in broad strokes, leaving later chapters to develop those ideas rigorously.

In one sense, the individualism–holism dispute is a small technical issue in the philosophy of the social sciences. Even the broader reductionist program may seem merely an issue in the philosophy of science. Yet lurking behind these apparently narrow questions lie much bigger issues.They run the gamut from practical questions about the practice of science to moral and political disputes about the nature of freedom and responsibility to large philosophical debates about rationality and knowledge. Later chapters often focus on the more technical philosophical issues. Nonetheless, it is important not to lose sight of what is ultimately at stake; the real motive forces driving the debate go far beyond the intricacies of reduction and supervenience in the social sciences. I want to describe those motives before outlining the more technical philosophical issues.

Moral issues are not far from the surface in the individualism–holism debate. Individualists explicitly worry that giving social entities like classes or institutions a fundamental role will rob humans of freedom, dignity, and responsibility. Individualism also supports distinctive views about the place of government policy and the nature of social justice. Holists locate the cause of poverty, crime, and the distribution of income in large social forces; individualists look for causes in the traits and preferences of individuals. Quite naturally, individualists are skeptical that governments can solve what they see as individual problems, while holists are likely to demur. Individualists are likewise naturally attracted to accounts of justice emphasizing individual preferences, contributions, merits, and rights; giving social structure an essential place raises doubts about that program.

Large issues about the nature of rationality and the self are also not far away. The Cartesian picture of the isolated epistemic agent, assessing claims to knowledge internally and independent of others, runs deep in the Western intellectual tradition. It remains the dominant presupposition in mainstream analytic epistemology and philosophy of science. It undergirds the theory of action that rational-choice explanations embody. So doubts about individualism indirectly raise these much larger issues, something I will argue in chapter 7.

Individualism represents the reductionist program applied to the social sciences. That reductionist program itself has widespread practical implications. Philosophers of science have often denied that theses about theory reduction entail anything in particular about how science should proceed. I think they are wrong. The obstacles that make theory reduction unlikely make reductionist heuristics suspect, for they are likely to see unity where there is diversity and diversity where there is unity—as I shall illustrate in chapters 4 and 5. Moreover, the reductionist program seldom stops at saying what *can* be done; it goes on to claim that integrating the special sciences—like biology, psychology, and sociology—is essential to their scientific adequacy. Behaviorism, genetic determinism, sociobiology, and many other eliminativist programs in the special sciences draw strength and inspiration from this reductionist credo.

Aside from these broad issues, the individualism–holism dispute and the reductionism it represents connect to numerous other important questions in philosophy. The nature of explanation, the relation between moral and nonmoral categories, the proper formulation of physicalism, the presuppositions of social contract arguments, and the place of folk psychology in a scientific world view are just a few examples.

Individualism and reductionism are not a single, cohesive doctrine; many different, logically independent theses are involved. Much confusion and bad argumentation comes from ignoring these differences. Later chapters will sort out these issues with considerable care. Here I provide a rough overview with two goals in mind: to identify from the start some widespread but fallacious moves from ontological claims to explanatory ones and to argue that theory reduction as traditionally understood plays an absolutely central role, in constrast to those contemporary reductionists who defend their views by emptying the idea of all content.

The reductionist program in general involves claims at least about ontology, theory reduction, explanation, confirmation, and heuristics. The ontological claims may assert, for example, that everything is composed of and dependent upon the physical or, more generally, that the entities of high-level theories like biology or psychology are composed of and determined by the entities described by lower-level theories such as chemistry and neuroscience. Claims about theory reduction argue that one theory can in principle do all the explanatory work of another; the classic reductionist program then holds that ultimately physics can do all the explanatory work that the special sciences—such as biology or psychology—do. Though claims about theory reduction are theses about explanation, many reductionists claim that even if reduction is not possible, lower-level theories have an explanatory primacy. Lower-level mechanisms, for example, may be essential to any successful explanation. Theses about confirmation are likewise common. Sometimes the claim is that no nonreducible theory can be well confirmed; at other times the claim is that lower-level mechanisms are essential to confirm high-level theories. Finally, reductionists offer advice about scientific discovery: the best way to produce good science is to seek reductions, to seek mechanisms, and so on.

Individualism is a specific application of these doctrines. Individualist ontological claims include the idea that society is solely composed of and dependent upon individuals. The individualist reductionist thesis is that all confirmed macrolevel, holistic explanations in terms of social entities can be captured by a theory referring only to individuals.

Claims about explanation include the idea that macrosociological explanations are inadequate without individualist microfoundations or that a complete account of individual behavior fully explains all we want to know about the social world. Assertions about confirmation include the idea that all evidence must be about individuals and that no theory is confirmed without individualist mechanisms. And, of course, individualists often claim that looking for explanations in purely individualist terms best promotes scientific progress.

How do we evaluate these theses? Both individualists and holists often claim that their position is some kind of conceptual truth—that more or less a priori, perfectly general considerations show the other side inevitably flawed. I will argue that this approach is misguided. The real issues are ultimately empirical ones about how the world works.

There is one general conceptual argument that is so commonly offered that I want to address it directly here, even though later chapters will point out variants of the same reasoning. Earlier I said that the various reductionist theses were logically independent. However, there is a widespread belief that they are not; in particular, it is common to assume that the ontological claims entail the reductionist and explanatory ones. The alleged connection is this: from the fact that some entity W is composed of and dependent upon its parts $P_1 \ldots P_n$, it follows that theories of W are reducible to theories of $P_1 \ldots P_n$. The inference occurs over and over again in the literature on methodological individualism. It is rife in discussions of psychology and neurobiology. This inference is so widespread that it may well be a candidate for an innate cognitive disability of the sort described by Amos Tversky and D. Kahneman (1982).

The inference should look suspicious simply because it claims an entailment between an ontological fact about the way the world is to an epistemological fact about what real human agents can know. That problem aside, there are obvious counterexamples as well. Here are some examples: Suppose that every program state of a computer—doing a binary sort, for example—is brought about by and dependent upon machine states. In other words, the program state just is a complex combination of electrical charges in the computer. Nonetheless, it does not follow that we can reduce the kind "doing a binary search" to some physical description of the computer. To reduce one to the other would require us to find a physical description that captured all and only binary searches. Nothing about the fact that each computational state is a physical state guarantees that we will ever find such a physical description; there may be, for example, indefinitely many ways to build a machine that does binary searches. We could develop similar examples using commonsense notions like "chair" and its physical realization or scientific notions like "fitness" and its physical embodiments.

That *W* is composed of *P* simply says nothing by itself about the ability of humans to explain or express theories of *W* in terms of theories of *P*. *Perhaps* God can legitimately draw such inferences. Yet, any reasonable notion of reducibility must assume that human agents have something less than divine powers.

This fallacious inference cuts deep because claims about theory reduction are fundamental. Many would deny this claim, and so I turn now briefly to (1) argue that claims about theory reduction underlie other reductionist theses and (2) defend the standard notion of theory reduction I employ against various attempts to substitute a much looser notion. First, the central place of theory reduction: theses about explanation, confirmation, and heuristics are much less plausible if theory reduction fails. Reductionists like to claim that lower-level theories can explain everything that needs explaining. Yet if higher-level theories describe causal patterns that cannot be derived from the appropriate lower-level theory, then there are important causal explanations the reductionists cannot give. Similarly, if theory reduction is beyond the pale, then the heuristic "seek reductions for good science" is misguided as is the demand that all evidence be expressed in lower-level terms. Claims about theory reduction are central.

I shall explain with care what theory reduction involves in the chapters that follow. However, the basic idea is simple and intuitive: one theory reduces another when it can do all the explanatory work of the reduced theory. To do so, it must be able to capture the basic categories and causal patterns identified by the theory to be reduced. It is a deep error, I think, to ignore this root notion of reduction, as many try to do in the process of defending the reductionist picture. In particular, it is a mistake to assume that molecular biology has surely reduced genetics or that statistical mechanics has surely reduced thermodynamics and then to argue any notion of reduction that says otherwise is misguided (see Hull 1974; Bickle 1995). Scientists use the word "reductionist" for all kinds of different rhetorical purposes and with numerous meanings—meanings they are often unclear about themselves. If there were good reasons to think that molecular biology and statistical mechanics could not do all the explanatory work of their higher-level counterparts (and there is), then whatever they have achieved, it is not reduction. To claim reduction while admitting explanatory incompleteness is to make the issue a trivial semantic one. It is not.

This trivializing move does have an understandable cause. Molecular biology and statistical mechanics have done something important for the unity of science. But that unity is not the reductionist one where in the end particle physics explains all. Instead, a look at real scientific practice will show that the sciences are unified by much more subtle routes. The basic tenets of the nonreductive unity I shall defend are as follows.

1. The special sciences such as biology, psychology, and economics are generally not reducible *as explanatory theories* to their lower-level counterparts.
2. This explanatory independence is consistent with an ontological dependence—biological, psychological, and social entities are composed of physical entities and dependent upon them.
3. Scientific unity comes from integrating the special sciences with their lower-level counterparts—by using one to test the other, by identifying the place of lower-level mechanisms and higher-level contexts, by using one to develop explanatory constraints for the other, and so on.
4. Scientific unity in real science comes from producing interlevel theories, ones that involve explanations from diverse areas connected in the ways mentioned above.

This picture of nonreductive unity applies to the individualism issue as well. It would argue for leaving explanations in terms of social entities an essential place while recognizing that social processes do not go on independently of individual behavior. It would argue that mechanisms tied to individual behavior have an important place in a complete story, but assert that the complete story even about individuals will be multilevel—in short, partly social.

Why has this (fairly obvious) picture of scientific unity been missed? Why have social scientists, in particular, persisted in the reductionist tradition so steadily? The answer is, of course, complicated. But I want to suggest several partial causes, for it will help motivate the arguments that follow. Earlier I jokingly ventured one explanation: innate cognitive illusions. The fallacious inference from "*W* is composed of *P*" to "*W* is fully explanable in terms of *P*" seems irresistible. However, I doubt that the roots are fully genetic. Among social scientists, the error results in large part because they have misunderstood what the natural sciences are doing. No doubt that misunderstanding itself has diverse causes, both large-scale social causes and much more particular causes rooted in the stories natural scientists tell about themselves. But I want first to push a little further the "error hypothesis"—the hypothesis that the appeal of individualism rests on a misunderstanding of how the natural sciences progress. I think there is good evidence that social scientists have misread both the biological and physical sciences.

Chapter 4 will argue that modern molecular biology has not reduced biology to chemistry but has produced instead an interlevel account that leaves the biological an essential place. I defend that thesis with many contemporary examples. Yet I think we could draw the same antireductionist moral from the history of biology as well. The usual simple story is that modern biology progressed by throwing off metaphysical views such as vitalism and teleology and replacing them with the reductionist strategy of deducing all of biology from the underlying molecular processes. Vitalism and teleology were infused with romanticist and religious ideology; only by adopting the materialist worldview of science with its reductive explanations did biology move into the modern era.

The real story is not so neat (see Lenoir 1982). Many nineteenth-century biologists labeled "vitalist" were not motivated by romanticism or religion. Those who defended teleology, such as Bergman and Leuckhart, did not oppose materialism. They did, however, oppose overly simplistic materialist approaches that ignored the role of biological organization and structure. For them, biological structure set the boundary conditions for chemical processes. I will argue in many places in the chapters ahead that some such fact confronts reductionism *in general*; individualism, for example, frequently fails because it presupposes rather than eliminates background social structure.

Similarly, von Baer, a leading nineteenth-century physiologist who was generally labeled a vitalist, criticized Darwin. Yet he did so not because of some mystical teleology or because of opposition to evolution. He simply thought that Darwin overemphasized the place of natural selection on small variations and of gradual species change—a complaint still advanced today. Moreover, there is good evidence that these movements played important roles in the advancement of biology. By emphasizing biological organization instead of the frequently crude biochemical explanations of time, they allowed modern biology to progress as a discipline in ways that strict adherence to the reductionist program precluded (see Allen 1975). So reductionism has looked so persuasive to social scientists because they have misunderstood the biology and its history.

Social scientists have similarly misunderstood physics. The reduction of thermodynamics to statistical mechanics is usually presented as a paradigm of reduction. Temperature is equated with mean kinetic energy and the gas laws are derived from Newton's laws applied to individual molecules. I don't doubt this reduction goes through more or less. Yet the gas laws far from exhaust thermodynamics. As Sklar (1993) has shown in careful detail, the entropy concept of the second law—which is at the heart of thermodynamics—has no reasonable correlate in statistical mechanics. We need knowledge from thermodynamics to choose the basic probability posits of statistical mechanics; in statistical mechanical terms, the time asymmetry and the approach to equilibrium have no inherent rationale. Thermodynamics thus retains an essential role in setting the boundary conditions for statistical mechanics. Once again, lower-level accounts presuppose structure identifiable only at a higher level, in this case thermodynamics. So the development of statistical mechanics does not unequivocally support the standard picture of progress through reduction, contrary to the common perception that it does.

So reductionism in the social sciences has been based in part on misunderstanding the natural sciences. It has also resulted from a faulty understanding of the social sciences—in particular, of their history. Modern-day individualists often read their methodological norms into the history of their discipline, making it seem as if individualism were a timeless methodological rule that lies behind the progressive success of their field. I want to illustrate one example of this Whigish strategy—namely, in economics. The stories modern-day economists tell themselves about their past history disguises the fact that individualism is a contingent and contentious doctrine. Moreover, the discipline of economics carries much weight inside the social sciences; other social scientists have bought the line that success in economics rests on its individualism. Thus this bit of history has large ramifications.

Joseph Schumpeter's *History of Economic Analysis* is a classic account that was published just as neoclassical economics and its individualism was solidifying its position as the dominant paradigm in the 1950s. Schumpeter himself was an economist with a professed interest in the historical and social side of economic processes. So when Schumpeter—who apparently coined the term "methodological individualism"—found individualism to be part of the timeless methodology of good economics, the message was particularly influential.

Schumpeter believed that economic analysis provided a batch of purely formal methods—a "tool box" in his own words—that could be applied to economic problems anywhere and everywhere. We shall see this idea again in chapter 7, where we see that individualism and other such methodological norms are often treated as purely formal constraints that depend on no domain-specific assumptions. For Schumpeter, the economist's tool box was one that was present in economics' founding thinkers and one that had seen a steady refining through the discipline's development. Schumpeter denied that Jevons and Walras produced a "revolution" in economics, claiming that they "visualized the economic process much as had J. S. Mill or even A. Smith" (1953, 892). Smith and the classicals, like the modern neoclassical picture, took a theory of prices to be the fundamental task of economics (1953, 911). General equilibrium theory was there all along in Smith and the other classical economists, though in rudimentary form (918). So the methodological individualism of modern neoclassical economics was part and parcel of economics from the beginning and helped explain its success.

Schumpeter's story, however, distorts the real history. There was a kind of individualism in the classical economists, but only in its weakest form—in the idea that large-scale processes and individual behavior need to be tied together. As we will see in chapter 3, that thesis is far from the radical views of modern-day individualists. It is compatible with a picture of social theory that gives institutions and social structure an essential, ineliminable role; it is compatible with explanations of individual behavior in social terms.

Moreover, this is exactly how the classical economists proceeded. Social structure and institutions were built in from the start. For Smith, Ricardo, and Marx, perhaps the fundamental question of economics was how to explain the distribution of income between social classes (see Dobb 1973). For the classicals, a theory of prices was not, in contrast to Schumpeter, the first question to answer. A theory of distribution was not simply a theory of prices, and their interest in the latter was secondary to their interest in large-scale dynamics. Nor was Adam Smith's invisible hand just general equilibrium theory struggling to break free. For Smith, the invisible hand idea was, above all, the claim that individual actions had unintended consequences that exhibited regularities; he explicitly acknowledged that it depended upon background social institutions, customs, and the like (see Gordon 1991). It was not the claim that economists can explain everything in individualist terms. Modern-day methodological individualism is thus not an inevitable part of the economists tool box.

So another historical mistake—this time a mistaken self conception—lies behind individualism's appeal. Of course, my histories are extremely cursory; moreover, these genetic arguments are not conclusive, for individualism might be a reasonable doctrine despite its confused origins. Yet seeing the real origins is an opening wedge to criticism, for much of individualism's motivation comes from its alleged ties to supposedly successful reductionist programs elsewhere.

Let me turn finally to a brief overview of the arguments to come. Chapters 2 and 3 target individualism. I work hard to sort out individualist theses, for failure to do so is a major obstacle to progress. I argue that individualism cannot be decided on conceptual grounds but ultimately is an empirical issue. Chapters 2 and 3 both try to say what those empirical issues are and to provide some reason for thinking the evidence favors the holist. The fact that social processes can be brought about in diverse ways and that individualist explanations presuppose background structure argue against reductionist versions of individualism. Ontological individualism has some plausible formulations, but they do nothing to support claims about reduction, explanation, and method. Good science can proceed without providing mechanisms and the same holds for social science; at most, individualist mechanisms are needed if we want the full story about causes. Yet nonindividualist theories likewise have an essential place, for they pick out causal patterns not capturable at the individualist level.

Chapters 4, 5, and 6 turn to defend a nonreductionist picture of science in general. Chapter 4 looks in detail at what biochemistry has done for biology. I argue that it has done much, but it has not replaced biological explanations. Even at the most molecular level, nonbiochemical categories have an essential place. Natural selection does not care about mechanisms, only about outcomes. So it is no surprise if biological systems can achieve the same end by diverse underlying processes. Similarly, biological structure is constantly presupposed rather than eliminated in explaining how molecular processes work. I argue that contemporary molecular biology is thus a paradigm of nonreductivist unity—it ties the biological to the chemical while leaving the biological an essential role.

Chapter 5 then turns to deflect objections to the nonreductivist approach I advocate. Most of the objections claim an inconsistency—between nonreductionism and the ontological commitments that wholes are constituted from their parts and supervene upon them. Some argue that supervenience guarantees reduction. Others argue that supervenience at least entails that lower-level accounts can fully explain. Still others argue that these ontological commitments leave no room for causation at any level beyond the physical; nonreductive unity empties the special sciences of any claims to capture real causal processes. I argue that these objections all fail, in large because they are unable to see the difference between reductionist claims about explanation and theory and reductionist claims about ontology.

Chapters 6 and 7 return to individualism, particularly in economics. One version of individualism that gets little discussion in chapters 2 and 3 concerns the claim that individualist theories provide the best explanation. Chapter 6 investigates this idea in detail, using neoclassical economics as a concrete study. I argue that inference to the best explanation depends on substantive, contingent empirical claims to do its work. I trace out those substantive empirical claims in the case of neoclassical theory. In the end I show that neoclassical theory and the individualism it embodies cannot be defended on explanatory grounds, for the assumptions it makes about explanation are quite empirically suspect.

Chapter 7 looks at individualist assumptions in accounts of rationality. My arguments here are a bit fast and furious, for my goal is to make a prima facie case about some very large issues. My basic thesis is that we cannot capture what is important about rationality if we treat rationality as a trait of individuals. First, I discuss individualist assumptions in traditional epistemology. A rational community need not be a community of rational individuals, and a community of rational individuals may not be rational. These conclusions commit no ontological sins—they do not commit us to group minds—and are moreover supported by the antifoundationalist trend in epistemology in ways that few have recognized. What holds for epistemic agents also holds for rational-choice theory, or so I argue. My claim is that rational-choice theory cannot capture collective phenomena *and* that it cannot explain individual behavior either if it proceeds individualistically. I discuss human capital theory, parts of game theory, and rational expectations to make these points. Thus, as I pointed out above, how we understand the unity of science is not just an idle issue in philosophy of science—it has important implications for how we understand and evaluate the social world and our conception of ourselves.

2

The Empirical Nature of the Individualism–Holism Dispute

In this chapter and the next I sketch a preliminary argument against methodological individualism. That argument will be deepened in chapters 4 and 5 as we look at reductionism in general, and then again in chapters 6 and 7 as we return to individualist approaches in economics and epistemology. My goal in this and the next chapter is not a decisive refutation of individualism. Rather, I seek above all to sort out the issues, for the individualism–holism debate has been long on rhetoric and short on clarity. I argue that the real issues are empirical; the dispute will not be settled by appeal to general conceptual considerations. This chapter describes in concrete terms some of the crucial empirical issues. Though I will hazard guesses about what the evidence will show, I save the more assertive arguments for chapters 3 and 7.

Sorting Out the Issues

Individualism and holism are nebulous doctrines with shifting meanings that are too often run together. So progress on these issues requires first carefully separating different theses involved and getting clear about their interconnections. I shall concentrate on individualist claims, taking holism to be their denial.

At the most general level, individualism involves at least claims about ontology, reduction, explanation, and confirmation or evidence. Common ontological claims are that social institutions or entities do not (1) exist separately from or (2) act independently of individuals. These two claims are probably the most common ontological versions of individualism and, as we will see below, are often used to support claims about reduction and explanation.

Another quite common version of individualism alleges that all social explanations can be reduced to theories about individuals. The term "reduction" is often used carelessly, but we can give it a precise sense that will focus the issue. For one theory to be reducible to another, we need a way of connecting the different terms or predicates of the two theories such that we can show that higher-level explanations can be in principle deduced from lower-level ones. While there is a debate over just what these connections require, it seems to me that if we want to keep the root notion behind reduction—namely, that one theory can in principle do all the explanatory work of another—then we need at least the following: (1) lawlike mappings from each social description to individualist ones that (2) allow us to deduce the true or well-confirmed explanations of social theory in such a way that (3) social explanations are completely replaced by individualist ones. The first requirement allows us to connect up terminology, which by the way need not be a connection of meaning. However, not just any lawlike connection will do; we need a connection that allows us to deduce explanations *and* allows us to do so in a way that shows we can in principle eliminate the reduced theory.

Related but separate from reductionist claims are claims about explanation. Here lurk many potential confusions. Individualists and holists sometimes make claims about full versus partial explanations; sometimes they make reference to individuals a necessary condition for explanation and other times only a sufficient condition for explanation; sometimes they require some reference to individuals and other times require reference *solely* to individuals. These various parameters of course need clarified, especially full and partial explanations. As you can guess, numerous different individualist claims can be constructed from these different parameters, among them:

1. Full explanation requires reference solely to individuals
2. Full explanation requires some reference to individuals
3. Purely individualist theories suffice to fully explain

The first thesis is a methodological norm, for it proposes a necessary condition for successful science. Thesis (2) does so as well, but its requirement is weaker. A common form of (2) is the demand for individualist mechanisms, a version of individualism espoused by Jon Elster (1989), for example. The third thesis leaves open the possibility that holist theories might be good science as well, and thus it is a weaker though still controversial form of individualism. Many other theses could be listed here, but it is these three that are most important for my purposes.

Finally, the individualism–holism dispute is frequently put in terms of evidence. Sometime individualists claim that all the evidence is evidence about individuals in some sense. A weaker version of that same idea is that no social account is well confirmed until we have evidence about individuals, particularly individualist mechanisms, even if there can be evidence about social entities. This is another way of reading Elster's demand for mechanisms. Other issues about evidence concern whether individualist theories are best confirmed or whether pursuing individualist accounts is the best heuristic for producing good theories in the social sciences.

These theses likewise call for further clarification. Particularly, the idea that individualist theories are best confirmed threatens incoherence, for individualist and holist accounts are at different levels and in different vocabularies and thus may not necessarily compete at all. Chapter 7 will look in some detail at the idea that individualism is the best explanation, and thus I save further clarification till then. In this chapter I will concentrate on claims about mechanisms and heuristics.

Thus, the individualism–holism dispute has many guises, and evaluating the debate really means evaluating a great many different theses. In what follows I focus on what I take to be the most important of these, beginning with the reductionist claim.

Individualism as a Reductionist Thesis

Both individualists and holists have thought that reducibility could be decided by appeal to some very general, almost commonsensical truths. From the individualist side we hear John Watkins (1973) saying that reducibility is entailed by some metaphysical truisms about individuals. On the other side, we hear holists claiming that reduction is ruled out by the obvious fact that individuals are causally influenced by social processes (Ruben 1985). These are what I called earlier "more or less a priori arguments"—they look for a quick decision by philosophical argument from some very general facts. Not surprisingly, I am unconvinced by either argument.

Individualists have thought social theory to be obviously reducible in principle because reduction appears guaranteed by several undeniable truths about social events. Watkins (1973, 179) cites two "metaphysical commonplaces" supporting reducibility: (1) "the ultimate constituents of the social world are individuals" and (2) "social events are brought about by people" or "it is people who determine history." These commonplaces, Watkins claims, have the "methodological implication that large-scale phenomena . . . should be explained in terms of the situations, dispositions, and beliefs of individuals" (179). This methodological injunction presupposes that reduction is in principle possible, and Watkins is explicitly committed to reducibility. Thus, reducibility allegedly follows from the fact that societies neither exist nor act independently of individuals. This reasoning occurs again and again in the methodological individualism literature.

Watkins's principles are obviously plausible, although simply identifying social institutions with sums of individuals may be somewhat controversial. However, even if we grant the individualist these principles, they do not of themselves entail reducibility. Let me tighten up Watkins's principles and then explain why they imply nothing about reducibility.

Watkins's metaphysical commonplaces say that social wholes are both composed of individuals and determined by their actions. Borrowing from Hellman and Thompson's (1975) discussion of physicalism, we can describe two principles here: an exhaustion and a determination principle. Individuals exhaust the social world in that every entity in the social realm is either an individual or a sum of such individuals. Individuals determine the social world in the intuitive sense that once all the relevant facts (expressed in the preferred individualist vocabulary) about individuals are set, then so too are all the facts about social entities, social events, and so on. Or, to put this idea in terms of supervenient properties, the social supervenes on the individual in the sense that any two social domains exactly alike in terms of the individuals and individual relations composing them would share the same social properties.

Both principles can be made more precise, but for our purposes they will do as formulated. The question at issue is whether the social realm can be exhausted and determined by individuals yet social theory be irreducible to individualist theory. As we saw, Watkins believes that these principles imply that social phenomena should be explained individualistically. Assuming that *should* implies *can*, Watkins is in effect claiming that supervenience entails reducibility. Mellor (1982, 70) also argues that social theory is reducible in the traditional sense and he likewise seems to think reducibility follows from supervenience: group laws "relate attributes *supervenient* on its member's actions and attitudes. The [group] law . . . must *therefore be derivable* from some true explanatory psychological theory" (my italics). Derivability is, of course, the goal of reduction as traditionally understood, and Mellor explicitly endorses this traditional account of reduction (1982, 51).

Both Watkins and Mellor are wrong, for determination and exhaustion do not *entail* reducibility of language, as I suggested briefly in the first chapter. The same mental state, the same program, the same camshaft, and the same social event can all be realized in very different configurations of composing elements. For example, it is logically possible that any number of different relations between individuals realize such social predicates as "peer group" or "bureaucracy," just as programs and camshafts can be embodied in indefinitely many physical materials and states. Thus, although each of the composing configurations determine and exhaust the events they realized, these multiple realizations could mean that there is no one configuration of composing elements coextensive with predicates of the reduced theory. However, reduction of language requires just such coextensionality. Exhaustion and determination of themselves, without further assumptions, do not show that reduction must be possible.

Moreover, even if we ignore this multiple-realizations problem, there are other potential obstacles that show reducibility is not inevitable despite Watkins's metaphysical truisms. Even if all the facts about individuals fix the social facts, that says nothing about how lower-level theories explain. If our accounts of individual behavior presuppose claims about social institutions, classes, and such, then those accounts—though about individuals and compatible with Watkins's metaphysical commonplaces—will not suffice for reduction. Why? Reductions replace one level of explanation for another. But if we presuppose facts about social variables in explaining individual behavior, the social will not have been eliminated. So from the fact that institutions are made up of individuals, nothing follows inevitably about the power of purely individualist theories to explain.

Much of methodological individualism's intuitive appeal comes from the fact that the exhaustion and determination principles are so plausible. We can now see that such appeal is ungrounded.

The holist argument against reducibility cited earlier—namely, that social entities causally influence individuals—fares no better. Some may doubt that we can coherently picture social wholes causing individuals. But let's grant the holists their claim (which I argue in chapter 5 poses no special problems). Still, it does not follow that theories about social institutions will be forever irreducible. Why? Individualists might well be able in turn to reduce the social-to-individual link to an individual–individual causal chain—just as we might explain the influence of a cell on specific molecules by explaining those cellular influences in molecular terms. Thus so-called downward causation does not show that reduction is impossible either.

Individualism as a reductionist thesis will not, in my view, be decided by arguments like these. If it could, then we could determine the course of all future science from a few very general facts. Of course, it would be amazing if we could do that. I suggest that our judgments about reducibility must be much less certain and much more empirical. Any claim about reducibility is a guarded judgment about how science will go given what we know about the world now. It is an empirical bet based on our best current knowledge. More specifically, it is a bet that potential obstacles to reduction will not be real. We saw two of those potential obstacles earlier when criticizing Watkins: social kind terms might be multiply realized by individual behavior and explanations of individual behavior may invoke social processes. Thus, some of the issues involved in reduction concern whether these possible obstacles are likely to be real-world obstacles. Do our best current explanations of *individual* behavior invoke social variables? Are social processes that we can identify—and note that there is no debate about reducibility if there is no social science to reduce—likely to be brought about by diverse collections of individual behaviors, at least given our best information now?

These are the kind of empirical judgments that I think are relevant to individualism as a reductionist thesis. Of course, these empirical questions are still highly abstract, and we can try to cash them out in more concrete terms. For example, we can ask questions like the following about the potential obstacle I labeled "presupposing social information": What assumptions do individualist accounts make about preferences? Are the preferences those accounts invoke likely to be "innate" or do they have social explanations? Are the effects of those preferences and the behavior they cause different in different social contexts? If so, appeal to social factors may be essential. Relatedly, we can ask whether explanations of individual behavior refer in some way to their group membership. For example, is the distribution of income between individuals a function of social class? of social class even when differences in education, intelligence, and other individual variables are controlled for? Questions like these help determine if so-called explanations in terms of individuals are not eliminated but instead presuppose explanations in social terms. Of course, even if individualist accounts do presuppose facts about social processes, those facts could themselves be reducible. But that just illustrates the empirical nature of the debate—for we would have to look again at the social process presupposed and ask about them whether they are multiply realizable and/or can be explained individualistically without residue.

For the multiple-realizations case, we would like to know if particular social processes, institutions, and so on can be found where the individuals involved have very different preferences, beliefs, and values. If so, we may have evidence that the institution involved may be brought about by a range of different individual behaviors. Alternatively, we can ask whether there are macrosociological processes that key in on the features of institutions and ignore individual-level detail. A positive answer is likewise evidence that social explanations are unlikely to be captured in individualist terms.

Let me mention some more specific empirical work that illustrates these kinds of issues, work that I think bodes ill for individualism as a reductionist thesis. One body of social research that I think is both relatively successful and positive evidence against reducibility is that of Hannan and Freeman (1989) on what they call "organizational ecology." Hannan and Freeman try to explain the kinds of organizations and their relative numbers by borrowing models from population biology. The basic idea is this: Organizations compete for resources (financial, political, etc.), so not every organization can survive. Organizations that adopt strategies that allow them to gain resources in their particular environment will survive or be successfully founded; those that do not will not. The end result will be a familiar sorting process like that in natural selection. Hannan and Freeman test specific versions of this general model for semiconductor firms, for fast food restaurants in a given locale, and for union organizations. As I have argued elsewhere (1996), their data are exemplary as social science research goes.

How does this work bear on the holism–individualism dispute? I think there is good reason to believe that the multiple-realizations problem will thwart attempts to reduce work like theirs. Like natural selection, which "cares" about phenotypes not genotypes, the social process they describe does not care about individual-level detail. It does not "see" such detail. As a result, there is much room for differences in implementation across organizations. For one, the basic strategies Hannan and Freeman describe—the generalist strategy, in which an organization focuses on a wide range of resources, and the specialist strategy, which focuses on a narrow range—will be embodied in different organizational forms depending on historical contingency, kinds of organizations, social environment, and so on. In short, there are many different ways for an organization to be a generalist or specialist. Moreover, any more-specific organizational form that embodies a specialist or generalist strategy may itself be compatible with diverse sets of individual attitudes, behaviors, and so on. So we have the prospect for multiple realizations at multiple levels. And we have an explanation for why that should be—the social process works through organizational outcomes, not specific individual behavior.

This suggests a generalization: social theories will be irreducible when they describe selection processes at the institutional level. If this is right, we can identify other areas where reduction is unlikely. An obvious test case in this regard concerns economic theories of the firm. Economists have a batch of predictions about how firms behave, and they sometimes defend those predictions on the grounds that firms that act otherwise will be eliminated by economic competition. Let's assume they are approximately right—that economic selection results in firms acting so as to maximize profits, equate marginal revenue with marginal cost, and the like. How would we expect such *corporate* behavior—the behavior of a social entity—to relate to individual behavior? I suggest that there may not be the neat relationship reduction requires. There may be numerous different strategies a firm can adopt in order to maximize profits.[1] For example, firms can have different internal structures, different market strategies, and different mixes of short-term versus long-term investment. Moreover, these strategies can be mixed in different ways. This diversity of organizational forms alone suggests that profit maximizing behavior will be realized in quite different sets of individual behaviors and attitudes. Furthermore, each component of the corporate strategy or structure for maximizing profits may arguably be brought about by diverse sets of individual behavior as well. So we can easily imagine how there might be good empirical evidence that accounts of corporate behavior are irreducible because of the multiple-realizations problem. Neoclassical economics tries to avoid such problems by treating corporations as if they were individuals. Yet, as we shall see later, these strategies are an ad hoc abandonment of the individualist program as well as technically flawed.

There is also evidence that the second obstacle to reduction—presupposing social explanations—is also likely to be real. Consider, for example, rational choice explanations, in particular the work of one of the founders of rational-choice theory, Gary Becker (1976, 1981).

Becker has made many creative and path-breaking applications of the rational-choice approach to social phenomena outside economics. In particular, he has argued that the rational-choice approach can explain employment discrimination and marriage patterns. In the former case, we assume individuals maximize, that white workers have a taste for discrimination, and that employment contracts are the result of market processes. With some further assumptions, Becker can show that wage differentials between races would result—without any attempt to discriminate on the part of employers. For the family, Becker is bolder yet: families result from a marriage market where individuals are maximizing their preferences—for kind of spouse, number and "quality" of children, amount of "caring," and so on—subject to the constraints imposed by the market.

Let's put aside the (big) question of whether Becker's work is confirmed and ask if his explanations support the individualist program, as the rational-choice approach is commonly thought to do. I do not think that it does. The reason is straightforward. Becker has to take as given the following: individual preferences for number of children and their "quality," individual preferences for traits of spouses, relative wage and property shares of males versus females, the taste for discrimination, and the relative shares of capital owned by the races.

All these "exogenous" factors are precisely ones that we would expect to explain *holistically*. Attitudes about children, about sex roles, and about racial prejudice seemingly have their roots in larger social structures—the family; the media; peer, work, and ethnic groups; religious affiliation; and governmental and judicial institutions. Relative property and income shares also depend on social factors—historical factors like slavery, patriarchy, and so on. In short, the exogenous factors all seem to be social. So Becker's work seems to presuppose social explanations rather than reduce them.

I would suggest that his work is typical in this regard—that if you look at rational choice accounts, they generally rely on a background of social institutions and processes. In fact, I will provide fairly detailed evidence for this claim in chapters 7 and 8. However, there is nothing inevitable in this, and in some domains rational choice accounts may be through and through individualist. But once again we see that assessing the individualism–holism dispute depends on contingent empirical detail. Critics might not be convinced by my assessment of the evidence, but I think it is over details like these that we must argue to progress.

The above work suggests another moral about the individualism–

holism dispute. Individualism as a reductionist thesis is a very strong thesis indeed; it only takes one case of good social science that is irreducible to refute it. But once we stop looking for all-or-nothing arguments that will decide the issue across the board, then there is room for interesting debate about a more-restricted reductionism. Individualism may be plausible in some domains or for some partial phenomena in those domains, even if it fails as a general thesis. Moreover, reductionist programs in the social sciences need not be individualist to be reductionist. For example, if economists claim to find microfoundation for a given macroeconomic process, that would be in the spirit of individualism—even if the reduction were able to explain in terms of corporate behavior, behavior that was not reducible to some individualist account. Again, this makes individualism an interesting claim only by forcing us to look at social research piece by piece.

Individualism and Explanation

Those with individualist sympathies, when confronted with the obstacles to reduction I have described, may remain unconvinced. They are likely to reply that every social event surely can be explained in individual terms, even if reduction fails, for we can describe in principle the behavior of every individual involved. This thesis is most interesting when it claims that social events can be *fully* explained, for the holist can allow that accounts in terms of individuals can partially explain. What should we make of this attempt to separate individualist explanation from reduction? Before we can answer that question we would ideally need a complete account of "explanation" and of "full" versus "partial" explanations. To avoid some tough problems about explanation, let me here use the idea that an explanation involves answering a question, the content of which may be determined by pragmatics, and that some central explanatory questions are questions about causation. Then one theory more fully explains than another when it answers more causal questions. Thus, we can make some *rough* sense of individualist claims about full and partial explanations this way without committing ourselves to the perhaps dubious notion of "complete explanation" and without assuming too much of substance about explanation itself. Put this way, the individualist thesis at issue now is whether individualist theories can answer all the causal questions that holist theories can—even if reduction fails.

When the individualist claims to fully explain every social event, there are two ways we can read that claim: as a claim to explain every social event *type* or as a claim to explain every event *token*—every specific event. Only the latter claim is of interest here, for if individualists had a way of capturing all social kinds, then they would have the bridge laws necessary for reduction. However, our interest now is in explanatory claims that might be true even if reduction fails. So the interesting individualist thesis here is that each social event can be explained case by case, and that such case-by-case explanations answer all the causal questions holist theories can.

As we shall see in the next two chapters, at least two problems potentially confront this individualist thesis: (1) Explanations case by case in terms of individual behavior may not be fully individualistic. That is, they may explain individual behavior partly in social terms. We saw this potential obstacle to reduction above, and it is also an obstacle to case-by-case explanation, for it belies the individualist claim to explain only in individualist terms. (2) A second potential obstacle is that important questions may go unanswered if the individualist ignores the social level. How is this so? An individualist account, without reduction, has no way to capture macrosociological kinds—in short, no way to capture patterns identified at the macro, or sociological, level. Of course, in some sense those patterns are there in a complete account of individual behavior. But they are there—if reduction fails—only as gerrymandered nominalist sums of individualist causes. From the individualist perspective, they do not look like kinds at all.

Again, whether these potential obstacles are real can only be decided empirically. For the first obstacle, the issues are the same as in the reduction case: how far we can get in explaining individual behavior in purely individualist terms. Let me give another example to flesh out the kinds of issues involved here. The ultimate goal for general equilibrium theories in economics is to take initial facts about individuals and then fully explain market phenomena. In principle, such accounts hope to explain all economic events, including large-scale events such as business cycles. Now a well-confirmed general equilibrium theory with all the facts about individuals might still fail to explain business-cycle events in fully individualist terms. How? By, for example, explaining individual economic behavior—say, of individual decisions on saving and hours of work—by reference to macroeconomic variables. In the case of savings decisions, those variables might include the interest rate or rates. Interest rates are, of course, macrolevel variables. So a full explanation in individualist terms would not be forthcoming. Whether this sort of problem is real in economics depends on the course of empirical inquiry. But I should note that most current econometric work on individual savings behavior and the like does indeed invoke macroeconomic variables.[2] So given what we know now, that work does not seem to strongly support the individualist position.

The second potential problem for case-by-case individualist explanations—namely, they miss macrosociological causal processes—raises a variety of questions. One fundamental question is ultimately whether there are causal regularities at all to be discovered at the social level. If there are no regularities to be found or none that can be captured by social science in something like its present form, then individualism cannot be faulted for not capturing social kinds. Thus, the individualism–holism debate assumes that the social sciences are not fatally flawed as science. This presumption involves numerous issues beyond my purview here—issues such as the probative force of nonexperimental evidence and the scientific status of ceteris paribus generalizations (which are rife in social science).

Once again I doubt these questions can be answered in the abstract; it will take a careful look at actual social research to explore them. In particular, I do not think we can decide whether the social sciences can identify macrosociological patterns by appealing to some general facts about causation. Critics sometimes claim that social entities cannot stand in causal relations because they are not the appropriate kinds of things. Yet surely we do not want to say that no aggregate, or state of an aggregate to speak more accurately, can be a true cause, for our ordinary causal claims about medium-sized objects are also about aggregates. There may well be some kinds of aggregates that may make little sense as causes; averages come to mind as an example.[3] And I think there are some interesting puzzles about causation when we are dealing with complex entities. But I doubt that such philosophical investigation alone is going to show there can be no causes at the sociological level.

Broaching the question of whether the social sciences explain at all leads into the next two individualist theses I want to consider—theses that make individualism a methodological imperative. Faced with irreducible social science or social patterns that cannot be captured case by case in individualist terms, individualists might shrug their shoulders and say so much the worse for social science. In short, one response to my discussion so far is to argue that individualism is a necessary prerequisite for good social science. So I now turn to individualism as a methodological imperative, starting with its most radical version—namely, that social theory must be reducible to be good science.

The traditional argument for this imperative goes like this: (1) good science must be unified with theories at more fundamentally levels, and (2) unification requires reduction, therefore (3) social theories must be reducible to individualist ones to be good science. Premise (1) I will discuss below in the context of a less-radical methodological imperative. Here I want to focus on the second premise. Again, I would argue that it is a broadly empirical claim—this time, an empirical claim about how science itself works. The individualist is claiming that the kind of unification found in paradigm cases of good science is that described by the reduction relation. Evaluating this claim then involves looking at both the history of science and at current science and asking how theories at different levels are interconnected. When different scientific domains are integrated, does that integration consist in showing that one domain is at least in principle deducible from the other?

Once we put the question this way, I think the empirical evidence clearly does not support this radical individualist methodological prescription. While I expect the reduction relation does describe the relation between some domains, the evidence suggests that there are other, more complex ways in which scientific domains may be integrated. As I will show in detail in chapter 4, cell biology and biochemistry are integrated not by the former being reduced to the latter but by the two having systematic connections of a different sort. The cell biology–biochemistry case suggests another route to scientific unity, one that involves at least the following:

1. Showing that the entities of one theory exhaust those of the other
2. Showing that two theories make no inconsistent presuppositions
3. Showing that the predicates of one theory supervene on those of another
4. Developing extensive evidential interdependencies
5. Developing extensive heuristical interdependencies
6. Developing extensive explanatory interdependencies

This unity-without-reducibility model I think is generally a more accurate account of how scientific theories at different levels relate. If this is right, then there is good evidence that unification does not require reduction and, thus, that reduction is no methodological imperative.

Are Mechanisms Necessary?

I want to turn now to a weaker methodological version of the individualism thesis: the claim that no social theory is adequate until we have individualist mechanisms. This claim is made by Jon Elster in a number of different publications. In Elster's writings, it is unclear exactly what the thesis is, whether it is a thesis about explanation or confirmation. Do we need to provide the individualist mechanisms to have an adequate *explanation*? Or do we need to know individualist underpinnings to have sufficient *evidence*? Since advocates sometimes suggest both, I will pursue both.

Elster argues for these claims as instances of general truth—namely, "to cite the cause is not enough: the causal mechanism must be provided" (1989, 4). As a general stricture on explanation or confirmation, this requirement is uncompelling. One initial problem is simply that "the" mechanism is of dubious sense, since we can describe mechanisms in more or less detail and at many different levels. If we take Elster's demand really seriously and ask for the most basic process connecting a particular cause and effect, then we will never confirm or explain until we have cited the facts about fundamental particles. That would, of course, make most all causal claims in science unconfirmed or nonexplanatory—including Elster's favored individualist claims.

So we need some less-ambitious methodological imperative if we want to make sense of the demand for mechanisms. I think there is no very interesting *general* moral about mechanisms. Instead, when and where mechanisms are needed will depend on context and substantive background knowledge. For the issue of confirmation, the key factors will be (1) how well confirmed our causal theory is, (2) what specific assumptions it makes about mechanisms, and (3) how well confirmed our theory is at the level of the mechanism. If we have good evidence for a causal claim and a very weak understanding of underlying processes, then the demand for mechanisms is not very compelling. This is especially true if our causal claim presupposes no specific mechanism or we can specify a range of plausible mechanisms that could underlie our causal process. Alternatively, mechanisms become more important as our causal claim makes specific assumptions about them and as our knowledge of events at the level of the mechanism becomes more certain.

Thus, in the abstract, there is no general demand for mechanisms in confirmation, again illustrating my theme that the individualism–holism disputes depend on concrete empirical detail. In the individualism–holism case, the key questions are how well confirmed we think holist accounts are, how well confirmed our accounts of individual level processes are, and how much individual-level detail our holist accounts presuppose. No single answer to these questions is likely to be forthcoming either. Sometimes the demand for individualist mechanisms will be plausible, yet sometimes holists can reasonably demand that alleged individualist processes be compatible with what we know about social processes.

Let me illustrate these points about mechanisms by previewing an example I shall discuss in detail in chapters 7 and 8: namely, the debate over microfoundations for macroeconomics.[4] Some economists reject any macroeconomic theory that has not been given clear microeconomic mechanisms. The most extreme version of this attitude is put forward by defenders of what is called "rational expectations."[5] Rational-expectations theories want a macroeconomics consistent with fully rational individual behavior, behavior so rational that individuals use—or act as if they use—all information about current and future economic variables that can be abstracted from market behavior and that they use the best economic theory available. How do we decide if this demand for individual-level mechanisms is reasonable? Given my earlier arguments, it will not do simply to claim that mechanisms are always needed as do some defenders of rational expectations. Instead, we have to ask how great our confidence is in the macroeconomic claims at issue, how much confidence we have in the rational expectations account of individual behavior, and, importantly, exactly what any given macroeconomic theory assumes about individuals.

So no universal ruling on the need for microfoundations is likely to be forthcoming. I doubt that rational expectations is a very well-confirmed theory of individual economic behavior; it rests on far too many unrealistic assumptions. Yet in some kinds of markets—markets where information is cheap, for example—the rational-expectations picture may be more reasonable. And while much macroeconomics may be weakly supported by the evidence, that does not preclude some specific macroeconomic causal claims from being relatively well confirmed. Furthermore, not all macroeconomic claims are alike in what they presuppose about individuals. Some versions of Keynesianism may assume individuals are systematically stupid. But that is not true of macroeconomics as a whole. Moreover, we might have a macroeconomics that avoided assumptions of systematic stupidity without going to the extreme of assuming individuals use all possible information and other such rational-expectations assumptions. These are again the kinds of specific empirical and theoretical issues that have to be investigated to determine when and where the demand for mechanisms is reasonable.

I have been discussing the demand for individualist mechanisms as a constraint on confirmation. What about this individualist thesis as a constraint on explanation? I doubt that supplying mechanisms is any general requirement, for the quite commonsensical reason cited earlier: we give explanations all the time that apparently succeed without citing underlying mechanisms. These cases include humdrum examples of explaining broken windows by flying balls while remaining ignorant of underlying processes as well as more exalted cases such as explaining evolution by appeal to natural selection while remaining ignorant of the underlying molecular genetic details. If we want to count these cases as explanations, then we cannot require that every causal claim cites underlying mechanisms to explain. We can cite causal processes without a full accounting of underlying events that bring them about.

A more motivated argument would require a developed picture of explanation, something beyond my purview here. However, let me just suggest the hypothesis that my account of when mechanisms are needed in confirmation can be extended to explanation as well: in short, adequate explanation will depend on contextual, domain-specific substantive assumptions (a thesis I illustrate in detail in chapter 6) and thus mechanisms may be required in some cases and not in others. Where would this hypothesis leave individualism as the thesis that holist explanations without individualist mechanisms are inadequate as explanations? My arguments suggest that such a thesis is not grounded in any general fact about explanation. Individualists will thus need to provide some specific fact about the social realm to support their conclusion. I know of no such fact, but even if one is adduced, it would again support my contention that the individualism–holism debate turns on specific empirical issues. The much more likely situation, it seems to me, is that in some contexts mechanisms will be needed for social explanation and in others they will not, and those contexts will be determined by substantive empirical information.

Individualism as a Heuristic

I want to turn finally to individualist theses about heuristics and scientific progress. By "heuristics" I have in mind claims less exclusive than the methodological norms just discussed. Those claims made assertions about what science must do to be good science. Yet there might still be interesting methodological versions of individualism that did not tell us what was necessary to do good science but only what sufficed—or at least sufficed as a rough rule of thumb. Put this way, individualism does not pretend to give us norms defining good science but only norms that are useful guides to scientific progress. Thus, the individualist might claim that "seek reductions of higher-level theories to lower-level ones" or "seek lower-level mechanisms" were rules that produce good science.

Though these heuristic claims are intended as empirical generalizations from the actual practice of science, evaluating them seems to leave plenty of work for philosophers. That is because (1) there are numerous conceptual issues to be clarified and (2) no empirical discipline is likely to pursue such issues. Consider first the conceptual issues. It has become fashionable lately to claim that methodological norms of all stripes—not just heuristic rules of thumb—are empirical claims, claims about what practices promote scientific goals.[6] While I share the sentiment behind that thesis, I am worried that it smooths over important ambiguities. "Scientific goals" hides numerous unclarities, for instance. First, if we are to use the history of science here, we have to ask whether we are concerned with scientific goals as currently practiced or with goals of the scientist under study. Second, even if we stick to current science, identifying "the" goals of science is a messy business. One obvious goal for evaluating methodologies is truth. But that is too easy. What we have access to is what we take to be true or perhaps well-confirmed beliefs given our background beliefs. Moreover, even "what we take to be true" is a gross oversimplification as "the" scientific goal. Presumably scientists want truths that explain, truths that lead to fruitful research problems, and truths that have pragmatic value. Thus the idea that we can evaluate reductionist heuristics by whether they promote scientific goals calls for a great deal of sorting out before we can even state the thesis or theses in question. And that sorting out, I would suggest in a Quinean spirit, ultimately requires coming up with a philosophy of science that has been tested against the practice of science itself. Obviously, there is enormous work to do here, work that will by default be largely done by philosophers.

Of course, once we clarify the individualist claims about heuristics, there would still be the enormous task of making any reasonable judgments about them based on the historical record of science and/or its current practices. Among the obvious complexities are deciding when the relevant scientific situations are sufficiently similar to warrant generalizing. That decision is necessary because a heuristic that is valuable with one set of empirical views about the world need not be so when those views are different. This is just one of many complexities that face any serious attempt to empirically evaluate methodologies.

My own suspicion is that there will not be any very powerful generalizations about heuristics. Some reductionist heuristics will work well in some contexts, fail in others. For example, though I suggested in chapter 1 that antireductionist approaches were useful in early nineteenth-century biology, they were perhaps a hindrance in the early twentieth century when organic chemistry was much more developed. Even if I am right in my skepticism about generalizations, investigation might still tell us something useful about the kinds of contexts where reductionist strategies fail and succeed. I suspect we could generate some worthwhile hypotheses about those contexts by looking at the obstacles to reduction discussed above—for example, that seeking reductions or proceeding in purely lower-level terms when multiple realizations are likely does not lead to progress, for in these contexts we will miss causal patterns at higher levels. But only a much more careful analysis could make such claims really plausible. Chapter 1 gave some sketchy evidence that reductionist heuristics can lead us astray; chapter 4 will offer further evidence as well.

3

Reduction, Explanation, and Individualism

The individualism–holism debate is an empirical one. The last chapter sketched what the empirical issues look like and made some suggestions about what the data may show. This chapter pushes the argument further, for there is a compelling case against the main individualist claims.

In what follows, I explicate, evaluate, and consider the connections among the following theses, each of which has been advocated by individualists or might be charitably attributed to them:

1. Social theories are reducible to individualist theories.
2. Any explanation of social phenomena must refer solely to individuals, their relations, their dispositions, and so on.
3. Any fully adequate explanation of social phenomena must refer solely to individuals, their relations, their dispositions, and so on.
4. Individualist theory suffices to fully explain social phenomena.
5. Individualist theory suffices to partially explain social phenomena.
6. Some reference to individuals is a necessary condition for any explanation of social phenomena.
7. Some reference to individuals is a necessary condition for any full explanation of social phenomena.

Taken together, (1)–(7) capture most variants of methodological individualism. I shall argue that (1), (2), (3), (4), and (6) above are highly implausible; that (5) is an open question; and that (7) is both plausible and more interesting than is initially apparent.

The Reductionist Thesis

Despite frequent denials by some adherents, methodological individualism involved from the start an assertion that sociological laws referring to social entities are reducible to theories referring only to individuals. For Watkins, methodological individualism is the thesis that no large-scale social events are explained "until we have deduced an account of them from statements about the dispositions, beliefs, resources, and interrelations of individuals" (Watkins 1973). Dore (1973, 77) read methodological individualism as the thesis that "all sociological laws are bound to be such as can ultimately be reduced to laws of individual behavior," and Martin (1972, 67–68) and Hempel (1966, 110) agreed with this reading. Mellor (1982) has defended versions of methodological individualism asserting that nonindividualist theories are reducible to individualist accounts. Similar claims are made by Elster (1985, 5).

Exactly what sort of reductionist claim is the individualist making? Initially, we can take the individualist to hold that (1) social theory referring to social entities and events is reducible to theory or theories referring only to individuals; (2) such reduction is perhaps not possible now but is in principle possible; and (3) reduction requires lawlike coextensionality between each primitive predicate of social theory and some predicate in the reducing theory that allow nonindividualist explanations to be replaced by individualist ones.

Some clarification of these requirements is in order. (1) While I cannot exhaustively define "social entity," obvious examples include classes, castes, nations, churches, bureaucracies, and peer groups. Social events and processes are then events and processes involving social entities; social theories are ones whose elementary or primitive terms refer to social entities, events, or processes. (2) Reduction is qualified with "in principle" for two reasons. First, the individualist need claim neither that our current theories support reduction (but only that some future, well-developed theories will be reducible), nor that such reduction must be carried out in complete detail, for it might be extremely cumbersome, albeit possible. Thus, the individualist only alleges that reduction can in principle be done. "In principle" must not, however, be taken too broadly for fear of trivializing the individualist's claim. The individualist thus must be asserting more than simply that it is logically possible for social science to be done completely individualistically. For our purposes, one theory is in principle reducible to another if it is reasonable to believe that our current theories could be extended or replaced with well-developed ones that allowed for reduction. (3)

Reduction does not require equivalence of meaning, but I shall argue that it does require biconditional bridge laws connecting primitive terminology of social theories with terminology of individualist theory.

The notion of reducibility employed here is a traditional one that is sometimes challenged (for example, Hull 1974). While I shall argue that those challenges are misguided, even if my arguments fail, methodological individualism as a traditional reductionist thesis still merits discussion because (1) past and present defenders of methodological individualism themselves assert that social theory is reducible in the traditional sense; (2) several important individualist claims about explanation (rather than reduction) ultimately turn out to presuppose a strict reductionism; and (3) some social scientists themselves explicitly advocate this version of individualism (for example, Homans 1974; Arrow 1968, 641); the attempt, for example, to provide microfoundations for macroeconomics is a reductionist program in the traditional sense (see Nelson 1984). Thus, thesis (1) as interpreted here retains an important place in any analysis of individualism.

There are three good reasons to think that reduction will fail on any likely development of social sciences: (1) multiple realizations of social events are likely; (2) individual actions have indefinitely many social descriptions depending on context; and (3) any workable individualist social theory will in all likelihood presuppose social facts. Each of these claims, if true, rules out reduction as defined here. The first two claims would prevent lawlike coextensionality of predicates, while the last prevents reduction even if coextensionality is established. Let me discuss each in turn.

We have already seen in the last chapter the problem that multiple realizations pose for reduction. Reduction requires an equivalence between social and individual terms. However, if one social term refers to an event or entity that can be realized by many different configurations of individuals, then no single individualist term will be forthcoming for any given social term. In short, one side of the required biconditional will fail.

Should we expect multiple realizations to be more than a logical possibility? I think so. Consider the following social terms: revolution, peer group, primary group, bureaucracy, and power elite. It seems clear that any number of different relations between individuals could realize the referent of these terms. This point becomes even more compelling when we realize that even particular institutions (tokens) persist through significant structural changes in the configurations realizing them. If necessary and sufficient conditions can be given for these social predicates, it will generally be by means of their function vis-à-vis other social institutions and events—much as psychological states might be defined in terms of their functional role in a cognitive system. For example, Cooley's (1956, 23–29) classic discussion of "primary groups" defines them in just this way. Social predicates may be definable, but generally not individualistically.

Individualists might grant multiple realizations but deny that they necessarily cause problems. Defenders of type materialism have argued that multiple realizations of mental states do not rule out reduction if the various physical states can be grouped into an appropriate kind (for example, Hill 1984). A like-minded individualist could argue that although social predicates have multiple instantiations in configurations of individuals, those individuals may be similar enough to constitute a kind that would be coextensive with the social predicate in question. While this possibility is real, in all likelihood it will not help the individualist. Similarity is relative to some respect, and the respect in which the different instantiating configurations of individuals may be similar will probably be *social* in nature. Individuals become a social group if they, individually and/or collectively, play a certain social role vis-à-vis each other, other *social* groups, larger *social* institutions, and so on. But appeal to these social facts generally reintroduces social predicates, thereby making reduction impossible. Thus multiple realizations in this case do undercut reduction.

Multiple realizations of social events make biconditionals between social and individual predicates unlikely, for each social event is not coextensive with any individual predicate. I want now to argue that a similar indeterminacy exists in the other direction; individual acts and relations, described solely in individualist terms, do not uniquely determine their social description, for they are context sensitive. For example, a worker who shoots her foreman may be involved in an act of terrorism in one case, religious conflict in another, and class conflict in yet a third. Such examples can be multiplied at will. They should cause us to look askance at any claim that individual behavior, described solely in individualist terms, will be uniquely correlated with some social description.[1] In fact, this sort of problem plagues reduction elsewhere—witness Searle's claim (1981) that intentional and functional states cannot be identified because a particular functional state can be correlated with different intentional states.

It may be objected that this latter problem arises only because the individual behavior in question is too narrowly specified. If we more broadly conceive individuated kinds of behavior by including the larger context, then we could eliminate any indeterminacy between individual and social events.[2] That may well be so, but invoking the "context" raises new problems. How is the context to be spelled out? The most obvious method of identifying the context is *socially*. The violent worker expressing religious hate is a member of the Catholic church, the foreman, a Protestant church. However, these contexts appeal explicitly to social institutions, and thus the individualist must look for an individualist description of them as well. Such a quest seems unpromising, for (1) these social predicates, as we saw before, probably will not have a unique description in individualist terms because of multiple realizations; and (2) the "surrounding" individual acts constituting the context may themselves not determine a unique social description. Appealing to the "context" thus threatens only to push the problem back a step instead of solving it.

The problem just raised leads us to yet another. The relevant context for describing an individual action often refers to a social role; two identical acts of physical violence may nonetheless be differentiated by the kinds of individuals involved (e.g., foreman, Catholic, etc.). Social theory naturally invokes such roles, since it must refer to types of individuals, not specific individuals themselves. However, many social role predicates are apparently unavailable to the reductionist, for they have an essential social content. Predicates such as teacher, employee, inmate, soldier, or citizen do refer to individuals, but it is reasonable to believe they implicitly involve social terminology as well. To have true statements employing these role predicates, we must also have true statements about social entities, for there are presumably no inmates without prisons, a judicial system, and laws and norms, and no teachers without schools. Applying any of these role predicates to someone seems to presuppose or entail a host of further facts about the social institutions that give them meaning. Elimination of social predicates thus becomes quite unlikely. This point is the gist of Mandelbaum's (1973) talk of "societal facts."

Social roles implicitly bring in social structure. Yet individualist accounts sometimes *explicitly* presuppose social structure as well. They do so by invoking an institutional background in the process of explaining individual behavior. Chapters 7 and 8 point out numerous examples of this practice in economics, where the distribution of income, preferences, norms, and so on are taken as primitive and unexplained. Below we shall see that similar problems confront a classic individualist approach in sociology.

Whether multiple realizations, context sensitivity, and social presuppositions rule out reduction is, of course, ultimately an empirical question. The considerations raised above do show, I think, the enormous difficulty facing any individualist–reductionist program. For example, future theories might dispense altogether with social role predicates that involve institutional membership; but given the social science we know now and its likely permutations, that prospect seems remote. Similarly, we might find purely individualist predicates that group multiple realizations or specify social contexts, but current social science, so far as I can see, is in no position to do so. We may not know a priori that social theory is irreducible to the individual any more than we know the same about mental predicates as they relate to the physical, but in both cases the prospects look dim. The burden of proof lies with the individualist.

A more convincing case for the irreducibility of social theory requires looking in detail at proposed individualist construals of social phenomena. Chapters 7 and 8 will do so for economics. Here I want to take up an alleged reduction in sociology. George Homans, an early advocate of methodological individualism, developed a partial theory of small groups, a theory he thought supported the individualist program. Homans attempted to expand behaviorist and rational preference theory to include group dynamics, and he claimed to be able to analyze group phenomena in terms of individual interaction motivated by "psychic profit" and the like. Among Homans's (1974, 166) results are conclusions such as (1) groups control their members by creating rewards they can withdraw and (2) groups ostracize deviants. Homans denies these statements make any essential commitment to social entities, for he can cite individual actions that are involved in "group rewarding."

As we shall see later in this chapter, while there is a sense in which Homans has given an individualist explanation, nothing he has said shows that group laws are reducible. The statement "social groups control their members by withdrawing rewards" faces all three blocks to reduction discussed above:

- *Multiple Realizations:* There will be innumerable social arrangements for rewarding and punishing. Material incentives, social esteem, religious salvation, or political power all can do the job, and do it in various ways. No one, set description of what individuals do to bring about "group rewarding" seems available. For example, we cannot group these various realizations simply by whether individuals are rewarded by their interactions, for not all rewarding interactions between individuals—for example, asocial behavior—realizes group rewarding.

- *Multiple Descriptions of Individual Behavior:* Not just any rewarding activity brings about "group rewarding"; whether a relation instantiates this predicate will depend on the social context. Some rewarding personal interaction will in some circumstances have nothing to do with group control or will even be counterproductive (for example, some sexual relations).
- *Implicit Social Content:* Explaining which personal interactions do count as "group rewarding" will naturally lead us to invoke the social roles that individuals play. Rewarding interactions between individuals will contribute to group control only if they bear some relation to group structure. But one natural way for that connection to be made is through social roles—interactions that reward as, say, teacher, inmate, or employee. However, these designations presuppose social truths—about schools, prisons, or corporations—as I argued before. In fact, other sociologists (Bates and Harvey 1975, 166) have raised criticisms of individualist small-group theory based in large part on this point and the preceding one. Thus, while individual interactions, perhaps even ones based on psychic profit, may bring about all social phenomena as Homans claims, reduction remains improbable.

We have seen so far that reduction is likely to fail because (1) social wholes and events have multiple realizations, (2) individual acts have multiple social descriptions, and (3) individualist accounts describing social roles have implicit social content. However, the first two claims are telling only on the assumption that reduction requires lawlike coextensionality between predicates. I want now to briefly consider some challenges to that assumption.

Individualists might grant both that social events have multiple realizations and that individual acts have multiple descriptions, and yet be unimpressed. Why not, we might ask, simply define social predicates disjunctively by listing all the individual configurations that realize some social event? While the details might be complex, so the reasoning goes, we can always in principle handle multiple realizations and descriptions by equating the predicate in question with the totality of different individual configurations realizing it. Mellor (1982, 53), for instance, apparently thinks such disjunctive definitions can in principle handle multiple relations. In his fashion the problems for reduction appear surmountable.

The appearance is deceptive. Disjunctive definitions as proposed here do not support reduction, for they will not preserve the truth value of counterfactuals contained in the social theory to be reduced. Given that social events have indefinitely many realizations, our disjunctive definition must proceed by listing all the actual individual configurations that have constituted the social entity or event in question. However, because such a definition refers only to actual realizations, it will be unable to handle counterfactuals, since they do not go beyond actual cases. Take, for instance, the claim from Marxist sociology that "if the French Revolution had failed, then there would still have been a bourgeois revolution in France." Defining "bourgeois revolution" disjunctively, we can list all the revolutions, described in individualist terms, which have been instances of the term "bourgeois revolution." The latter is thus replaced by "the French revolution or the English revolution or . . . ," with each of these particular revolutions described in individualist terms. But obviously the counterfactual claim under consideration cannot be true once reduced by means of this definition, since its truth would require that the English revolution or some other one occurred in France! (Note that it is the counterfactual nature of the higher-order statement—not its Marxist character—that causes the problem. Statements such as "if the stock market crash of 1929 had not taken place, there still would have been a depression in the near future," would also become false when translated.)

Such disjunctive definitions provide only accidental coextensionality; and as these examples illustrate, that does not provide an acceptable reduction. The individualist cannot avoid the problem of multiple realizations by this route.

If disjunctive definitions fail for reduction, we might try to deny as some have done that reduction requires biconditionals between predicates. I want to comment briefly on one such proposal.

Mellor (1982) has claimed that reduction demands only approximations relating reduced and reducing terminology and that approximations of this sort are readily available. He is wrong on both counts. A reduction, roughly put, takes place when one more fundamental or extensive theory can be shown to explain in principle everything that the reduced theory does. But, of course, to show that some social event is in x percent of the cases correlated with some type of individualist event is equally to admit that in 100-x percent of the cases individualist theory cannot do what social theory can. In short, as long as there are exceptions unexplained by individualist theory and explained by social theory, the former has failed to replace the latter. Approximations in such cases do not suffice for reduction.[3]

Furthermore, even if approximations do work for reduction, I see no reason to agree with Mellor that we already have many such approximations in the social sciences. Mellor (1982, 53) offers in support of his claim the following: "There are obvious general links between social and psychological phenomena: between language and perceptual ability, for example, and between economics and desires . . . enough laws may link psychology and sociology to reduce the latter to the former." While there may be such laws as Mellor cites, they have little to do with reduction. Laws may link phenomena without providing a reduction of one to the other. To use Mellor's example, language may influence perceptual ability in regular ways, but that hardly means we can define language in terms of those perceptual abilities. Similarly, the rate of inflation may be linked to people's desires for savings versus consumption, but it would be silly to think we can equate, for the purposes of reduction, inflation with those desires. Not just any link will do. (And even if we had the right approximations—between primitive predicates—we would still face the problem of implicit social content discussed above.)

Richardson (1979) has proposed the even stronger thesis that reduction requires no bridge laws whatsoever—neither biconditionals nor conditionals linking terminology are essential to reduction. Hull (1974) seems to support a similar position. Citing historical cases as evidence, these authors argue that reduction has been misunderstood by philosophers of science. Mendelian genetics, for example, has clearly been reduced to molecular genetics: we have molecular explanations even if no bridge laws are forthcoming. Therefore, they conclude, standard accounts of reduction must be given up. Because this thesis is essentially about explanation, I shall postpone discussing it until the next section, which considers explanatory versions of methodological individualism.

Explanation without Reduction

Rather than a thesis about reduction of theories, methodological individualism might be a thesis about explanation. Danto (1973, 328), for instance, reads methodological individualism as the claim that social events and entities "can only be explained by reference to [individuals]." Watkins (1973a) also insists that we have a "full" or "rock-bottom" explanation only when we have explained things solely in terms of individuals. This section accordingly looks at methodological individualism as the claim that individualist theories can explain the social world without loss.

"Can explain," of course, hides several different theses. In this section, I want to consider theses (2)–(5) on my original list, namely:

2. Any explanation of social phenomena must refer solely to individuals.
3. Any fully adequate explanation of social phenomena must refer solely to individuals.
4. Individualist theory suffices to fully explain social phenomena.
5. Individualist theory suffices to partially explain social phenomena.

Of these theses, (2) is clearly the strongest and (5) the weakest. Theses (2)–(4) are, I shall argue, false; (5) is true but only on its weakest interpretation.

Anyone advocating (2) is a radical individualist: he or she holds that no theory making any reference to social entities ever explains, even if the variables of the theory range for the most part over individuals. Radical views are not necessarily suspect, but this one surely is. It entails that nearly all sociology, anthropology, and even microeconomics—which refers not only to individuals but also to corporations, a social entity—are pseudoexplanations. No matter how suspicious one is of these "soft" sciences, it is hard to deny that they do on occasion explain. Economic laws relating supply, demand, and the behavior of firms, for example, are probably as well confirmed as much in, say, evolutionary biology and geology. Some social theories apparently do explain.

Theses (3) and (4) can be considered as a unit. Since thesis (4) claims only that social phenomena can be fully explained individualistically while (3) makes individualism a must for full explanation, a refutation of (4) will inter alia be a refutation of (3). However, the idea that individualist theory suffices to explain fully seems to me clearly implausible.

As we saw briefly in the last chapter, thesis (4) faces the following dilemma: either social phenomena are to be explained as types, in which case thesis (4) turns out to be nothing but an already rejected claim about theory reduction, or it claims that social tokens can be fully explained individualistically. Individualist explanations of the latter sort, however, are barely explanations at all and certainly not fully adequate as the individualist asserts (see chapter 5 for a detailed discussion of a more general form of this argument that applies to reductionist accounts in any domain, not just individualism).[4]

Consider the claim that every social *type* (for example, recessions) can be fully explained individualistically. Explanation is done by theories, and the individualist is now claiming that individualist theory can in principle do all the work of social theory. Thus, there is allegedly for each kind of social event or entity an explanation in solely individualist terms. However, the individualist is now committed to supplying an individualist equivalent for the types picked out by the social-kind terms. Therefore, this explanatory version turns out to be, or at least require, reduction of theories—in short, the strict reductionist thesis (1). That thesis, however, has already been shown implausible.

A second alternative is to explain social phenomena in effect case by case. While we might not be able to give a single individualist account of social event kinds, we could still explain the social realm completely in that we can explain every particular occurrence of social events, or so the individualist might reason. In what sense can we provide a full explanation by explaining social tokens one by one? I can see two possibilities. On the one hand, the individualist may think that social phenomena are fully explainable in that we can simply describe—for every particular social event—the individuals involved and their interrelations. Because social entities and events are realized in or supervenient upon individuals, there must be a description of that event referring only to individuals. We can presumably name and describe the particular individuals involved and their interrelations in solely individual terms—in physical terms if need be.

The second way in which particular events might be explainable in individualist terms is considerably more interesting. There might be individualist laws describing human behavior that apply to the particular behaviors realizing a specific social event, even though the predicates of those laws were not coextensive with ones in social theory. In other words, the configuration of individuals as a whole might fall under a social description, while the individualist laws, for example, the laws of psychology, might apply to individuals one by one in a way that allowed no equivalence of social and individual terms. Such a situation would be analogous, for example, to computer programs that potentially have infinitely many realizations: we may not be able to define programs in terms of machine states, but there are still laws to explain any particular realization in physical terms. Thus we seem to have here a real possibility for individualist explanation.

Neither of these two possibilities turns out to provide full explanations in individualist terms. In order to defend this claim, I need first to say something about what I take a full explanation to be. Following much recent work (van Fraassen 1980; Garfinkel 1981; Achinstein 1983), an explanation can be considered in part an answer to a why question. Questions are not answered simpliciter, but rather are given answers relative to a number of contextual parameters. The same question may be given different answers depending upon the contrast class involved ("Why did John die—in contrast to Bill or in contrast to recovering, for example?"). Furthermore, once we set the contrast class of possible answers, some restriction on kinds of answers must be specified—for example, the immediate cause, the structural cause, or the microdeterminants. This brief sketch suggests that the explanatory power of a theory can be evaluated along at least two dimensions: (1) the extent to which it can adequately answer any given, fully specified question, and (2) the number of relevant questions it can answer.[5] Most of the standard literature on explanation analyzes the first dimension. The second dimension appeals to our sense that a theory that cannot answer important questions is incomplete. Combining these two dimensions, we can thus say that a fully explanatory theory is at least one that can fully answer all relevant questions.

The above discussion is sketchy and will be developed in more detail in the next chapter. However, for my present purposes it will suffice to evaluate the individualist's claim to explain fully by proceeding case by case.

Individualist theory cannot provide full explanations of social phenomena by explaining only social tokens largely because such token explanations leave important questions unanswered. More specifically, for any social phenomenon or event *S*, two fundamental kinds of questions are the following:

1. Why did *S* occur, that is, what are the causal connections between the kind of social events (events involving social entities) preceding *S* and the onset of *S*?
2. Why does this kind of event (the kind to which *S* belongs) occur, that is, what other kinds of antecedent social events might bring about this kind of social event?

Both questions are the general form of a great many specific questions asked in social explanation and explanation in general. Question (1) could include questions about what kind of preceding changes in political, religious, intellectual, economic, class, or educational institutions causally influenced *S*. If functional explanations are not a species of causal explanations, then (1) could likewise include questions about the functional roles *S* plays vis-à-vis other social events and institutions. Question (2) seeks to place *S* in a more unified framework by citing the kinds of preceding social events that could cause the event but were not in fact actually present. Any theory of social phenomena that cannot answer these questions is prima facie not a complete explanation.

Individualist explanations of social tokens will not be able to answer questions falling under (1) and (2) for a simple reason: it has no way of referring to kinds of social events. Both (1) and (2) involve situating a particular social event in a web of causal laws connecting different social events. Causal laws, however, are about kinds of events, not event tokens. Thus the individualist who claims to give a full explanation—despite having no individualist terms coextensive with social kinds—must be in part wrong. Without tools to specify social-event kinds, questions like (1) and (2) cannot be answered. Any theory that fails in this regard cannot claim to be full or complete.

Thus the supervenience of the social upon the individual does not entail that social events can be fully explained individualistically as has been claimed (McDonald and Pettit 1981, 125). Supervenience ensures that we can describe what individuals did, for example, in bringing about the French Revolution and we might go on to invoke laws of psychology or other laws about individuals to say why they behaved as they did. Such a story would be explanatory, but it surely would fall short of being a full explanation. We would have no way to understand this event as a kind and, in particular, as a social kind. Consequently, we could not explain its causal connection to preceding changes in classes, religious institutions, and so on, of French society nor could we understand this revolution by relating it to other revolutions, other kinds of revolutions, and other political and economic transformations that have or might occur elsewhere. In short, many standard questions could not be answered.

The above argument against the explanatory thesis presupposes one crucial assumption: that social theory provides at least some successful explanations relating kinds of events. If social theory never got beyond piecemeal and isolated descriptions or classifications of social phenomena, then individualist explanation of social tokens would look relatively good. Social theory would not be able to answer questions (1)–(3) above and, consequently, we would be missing nothing by giving individualist explanations that referred only to social tokens. The individualist could thus defend the explanatory thesis roughly as Churchland

(1978) has defended eliminative materialism—by arguing that there is no successful theory to reduce.

What could justify such an eliminativism about all social theories—about any theory explaining in terms of social entities? We have no reason to think that an eliminativism of this sort follows from any *general* principles about good science. We saw in chapter 2 and will see again in chapter 4 that the demand for unity of the sciences is no help. Nonreducible theories can nonetheless be unified in various other ways with the rest of science; if all nonreducible sciences were candidates for elimination, then most science outside physics would be ruled nonexplanatory as well. We also saw in chapter 2 that there is no universal requirement that all good explanations have mechanisms. And surely there is nothing in principle wrong with theories that explain in terms of aggregates or complex wholes, for much of our best physics and biology does so as well. So social theories cannot be rejected on those grounds either.

Apparently any argument that all social theories are nonexplanatory must turn on something specific to those theories. Yet it is hard to see what such an argument would be (and its a priori nature should likewise make us suspicious). I will argue in chapter 6 that there is no good reason to think that individualist explanations are always the *best* explanation. I have also argued in great detail elsewhere (Kincaid 1996) that actual social research sometimes produces well-confirmed causal explanations at the social level. So the burden of proof is on the skeptic.

We can now also see what is wrong with Richardson (1979) and Hull's (1974) redefinition of reduction mentioned earlier. If one event token is realized in another, then we can in a weak sense explain the former in terms of the latter. But if (1) the higher-level theory really does provide explanations of social phenomena and (2) the predicates of those explanations have no equivalent in individualist theory, then the latter theory cannot claim to replace or supersede social theory, for there will be much that only the higher-level account may do.[6] In other words, if reduction fails, social theory will provide explanatory kinds and relations that cannot be found at the individual level. In fact, this conclusion holds generally for higher- and lower-order theories lacking coextensional predicates. Kitcher (1984), for example, has argued that molecular genetics has not reduced (in the sense of completely replaced) Mendelian genetics for similar reasons.

So we can now conclude that methodological individualism thesis (4)—that individualist theory suffices to fully explain—fails. Either this thesis turns out simply to be the reduction thesis if we are talking about types, or it tries unsuccessfully to fully explain case by case. But if (4) fails, then so does the claim that any fully explanatory theory must refer only to individuals. To deny that individualist theory can fully explain is equally to deny that only individualist theories fully explain.

Having rejected theses (2)–(4), we need finally to consider the weakest individualist claim for explanatory adequacy, namely, that individualist theory suffices to explain, even if not fully. It would seem that this anemic claim is guaranteed by supervenience. If the social supervenes on the individual, then at least every particular social event is determined by some relation between individuals. Any theory describing those relations will thus be a partial explanation of social phenomena. Although such an account will not answer questions concerning social events as kinds, it will nonetheless answer at least one sort of question—it will tell us why a particular event happened by citing the microevents that caused it.

There is reason, however, to question whether a purely individualist account of individuals and their relations does suffice to describe the facts realizing the social. The individualist version of supervenience holds that once all the facts about individuals, described solely in individualist terms, are set, then so too are all the facts about social phenomena. However, it is unclear to me whether this minimal set of facts suffices to determine all the social facts. If, however, individual facts do not determine the social facts, then citing the former will not give us a microexplanation of the latter.

There is at least one serious problem for the claim that all the individual facts determine the social facts. This individualist version of supervenience presupposes that there can be laws of human behavior that (1) are sufficiently strong to fully explain the individual behaviors making up any particular social event and (2) also make no reference to or presuppose facts about social entities. Are laws fulfilling both (1) and (2) likely to be developed? Obviously, we have nothing approaching them now, but this lack does not of itself rule out relevant future developments. However, an affirmative answer seems possible only on certain assumptions about human nature. If the basic patterns of human behavior depend significantly on social contexts—the institutions and culture one belongs to, the social roles one adopts—then any laws strong enough to explain will have to make reference to these social facts and thus will fail to be purely individualistic explanations. Of course, even if social contexts must be invoked, the possibility remains that those contexts could be cashed out individualistically. That move, however, threatens to reinstate the strict reductionist program. Whether human behavior can ultimately be explained without appeal to unreduced social facts is an empirical question. Affirming thesis (5) thus requires that question to be answered in the affirmative.

The Truth in Individualism

So far I have given little ground to individualism. Nonetheless, the last section did make a concession: individualist explanations are possible, even if not fully adequate. I want to make further concessions in this section and say what seems true and interesting in methodological individualism.

Think, for the moment, about the most radical of the holist metaphysicians, namely, Hegel. When Hegel claims history is the necessary progression of human civilization toward greater human freedom and self-awareness, what exactly is wrong with his claim? Aside from questions of confirmation (!), a major problem comes from the fact that we do not see, as Dray has argued (1964), how or why these large-scale patterns should come about. In contrast, when economists say that markets develop and influence their participants like an invisible hand, our misgivings diminish, or at least they should. What is the difference? Invisible hand explanations offer a *mechanism* based on the self-interested actions of individuals. It is this legitimate demand for this link between the micro and the macro that in part constitutes the intuitive appeal of individualism.

Let me formulate this demand more clearly. There are two things it does not involve. Above all, the need for an account in terms of individuals does not support any form of reduction considered above. Nothing in Smith's invisible hand, for example, implies that all social terms can be defined individualistically, for accounts of economic mechanisms may presuppose truth about social-level patterns.

Second, the need to refer to individuals does not warrant the claim that some reference to individuals is a necessary condition for successful explanation (methodological individualism thesis [6]). Earlier I argued that it was unreasonable to hold that every theory making some reference to social entities or social kinds failed to explain. It seems to me likewise true that purely social theories do explain or at least can in principle. Macroeconomic theories, for instance, are often formulated so that they refer only to social entities—nations, corporations, industries, households, or government institutions. I see no reason in principle to deny they explain simply because they do not refer directly to individuals.[7]

If reference to individuals is not necessary for social explanation, the individualist can much more plausibly claim that any social explanation that makes no reference to individuals, in particular to mechanisms involving individuals that bring about social events, has not given a complete or full explanation. We may give real explanations at the social level, but we have done so only partially.

The thesis—that some reference to individuals is a necessary condition for full explanation (methodological individualism thesis [7])—carries more punch than may be obvious at first glance. Presumably for every entity with composing elements, there is always something that can in principle be said about its microstructure that adds somewhat to our comprehension or, in our earlier terminology, answers further "why questions." However, in a great many cases, information about microstructure is far from essential. We can explain planetary motion, for example, despite the fact that we cannot cite the specific quantum-mechanical details that realize the planets. In this case information about microstructure is not needed to explain adequately. As we saw in the last chapter, there is no universal demand for mechanisms in explanation.

Thus, when the individualist asserts that explanation in the social realm requires looking at the microstructure—individuals and their relations—something nontrivial is being said. Earlier we distinguished two explanatory virtues of theories: how successfully they answer particular questions and the number of questions they answer. Individualists are asserting that questions concerning individuals have some kind of special importance. In my view, that importance is largely pragmatic. We find social explanations which make no reference to individuals incomplete because of the "closeness" of the social and individual levels (unlike the planetary and quantum-physical levels), because much of our information about social entities comes from observing what individuals do, because many social predicates (like "needs," "interests," and "goals" of institutions) apply equally and paradigmatically to individuals, because of our natural practical interest in how things relate to ourselves not just to, for example, "the middle class." Such factors lead us to hold that the important questions and the relevant kinds of answers for complete social explanation must make some reference to mechanisms involving individuals.

Individualists have not strictly separated this more-plausible claim from the previous, less-convincing versions of methodological individualism. Still, when Homans (1974, 384) demands that social explanation cite the "underlying mechanisms of human behavior" and Popper (1950, 291) says, "we should never be satisfied with an explanation in terms of so-called collectives," they are clearly endorsing something akin to the completeness thesis. Methodological individualism, understood this way, gains some support from reductionism elsewhere. Mendelian genetics apparently cannot be reduced to molecular genetics nor be completely replaced by it. Nonetheless, molecular genetics gives us what Kitcher (1984) calls an "explanatory extension" of classical genetics. An explanatory extension reveals the fine structure underlying the basic elements and processes of a theory (although not replacing or reducing that theory). Classical genetics remained in some sense incomplete until molecular biologists provided an explanatory extension. We can thus plausibly point to this example and others as lending credibility to the parallel individualistic claim.

So at heart the rational core of individualism is the demand that complete theories be *interlevel*: we should combine accounts in terms of social entities and structure with accounts of underlying processes involving individuals. This view is indeed plausible, for as I shall show in the next chapter on molecular biology, it is what real scientific unity comes to. But note that this truth in individualism leaves the standard individualist claims behind, for the social has an essential place, perhaps even in the explanation of individual behavior itself.

We have now discussed all the versions of methodological individualism listed in the introduction. As a claim about theory reduction, methodological individualism is highly implausible. It likewise fails when it restricts all explanation to the individualist level or even makes such reference only necessary for explanation. Much more plausible, however, is the individualist intuition that explanations that refer only to social entities remain incomplete.

4

Molecular Biology and the Unity of Science

So far I have argued against reductionism in its individualist guises, and I shall return to other versions of that doctrine in chapters 7 and 8. Now, however, I want to broaden the scope of my attack. Reductionism is implausible even in those cases that seem to be a paragon of reductionist methodology. In particular I want to argue that the enormous achievements of molecular biology do not support the reductionist program. I do so by looking in detail at recent work in cell biology and biochemistry and arguing that precisely the same general obstacles that defeat individualism are present for any proposed reduction of cell biology to biochemistry. Moreover, the real relation between cell biology and biochemistry exemplifies the sort of nonreductive unity of science I advocate. Thus, this chapter serves three functions: (1) it extends my general argument to reductionism in general and to its apparently best case; (2) it shows what nonreductive scientific unity looks like in practice; and (3) it further undermines the individualist program, for individualism can no longer appeal to the place of reduction in the natural sciences.

The section below briefly reviews what reduction requires and the potential obstacles to reduction. The section following it discusses in detail some exciting work in molecular biology—the signal hypothesis—and argues that this work is irreducible to biochemistry. Further research is then discussed that also supports the antireductionist position. After that I show how these conclusions impinge upon both the

---

Parts of this chapter were previously published in *Philosophy of Science* (1990, vol. 57, pp. 575–93) and appear here by permission of the publisher.

heuristics and experimental practice of molecular biology. Finally, the last section sketches a nonreductive account of unity and argues that molecular biology, properly understood, is a paradigm for such unity.

Requirements for and Obstacles to Reduction

One core notion behind "reducibility" has always been that one theory can be shown to do all the work of another—in the way that statistical mechanics allegedly can account for and explain the gas laws. By showing that the higher-level (or "reduced") theories are just special cases of more universal and fundamental lower-level theories, reduction provides explanatory and ontological simplicity and shows how the many sciences fit into one cohesive picture.[1] However, one theory cannot explain everything another can without some way of trying the two theories together. More specifically, reduced and reducing theories usually describe the world in different terms. To show that a higher-level theory is really just a special case of its lower-level counterpart, we need, on the standard account of reduction, bridge laws showing us how to equate each essential higher-level term or kind with a description in the reducing theory.

Given such bridge laws, the reductionist holds that lower-level theory plus a description of initial conditions will allow us to deduce a purely lower-level explanation of everything described by the relevant higher-level theory. Of course, the reductionist need not claim that such explanations have already been provided or that they ever will be. Rather, the reductionist thesis is one about what can be done in principle. Claiming reducibility is asserting that if lower-level theories were fully developed, they would suffice to derive all adequate higher-level explanations, assuming we removed all limitations of time and resources.

How are claims about reducibility to be evaluated? While philosophers and biologists have sometimes given more or less a priori or conceptual arguments, I have argued that reducibility is ultimately an empirical claim. It is a judgment, based on our best current evidence, about the relations between theories and their foreseeable permutations.

In what follows I will focus on three general kinds of obstacles to reduction:

1. *The problem of multiple realizations.* Reducing one theory to another requires that we have some lawlike way to connect the vocabulary of the two theories. If, however, the kinds of events

described by the higher-level theory are brought about by indefinitely many different kinds of lower-level entities, then such a lawful connection may be lacking. As a result, we will have no way of deducing higher-level explanations from lower-level ones. Reduction will thus fail.

2. *The problem of context sensitivity.* Reduction as standardly conceived requires that we identify a determinate connection between lower- and higher-level descriptions. However, it might well be that events described in lower-level terms are not uniquely correlated with some higher-level description, depending on the context. If that happens, then it becomes difficult to replace higher-level descriptions with lower-level counterparts—since the latter vary in their higher-level significance.
3. *The problem of presupposing higher-level explanations.* This problem has been largely ignored by advocates and defenders of biological reduction, although it has been extensively discussed in similar debates in the social sciences. If the goal of reduction is to completely replace higher-level explanations by those at the lower-level, then it is essential that lower-level explanations proceed in entirely lower-level terms. However, it is quite possible for reductionists to provide reductive bridge laws and yet fail to reduce, by explicitly or implicitly using or presupposing higher-level explanations or descriptions in the process. Defenders of reducibility such as Schaffner (1993) completely ignore this potential problem, yet we will see that it surfaces frequently in the biological realm.

A reducing theory may presuppose higher-level facts or truths in its own explanations in at least two related ways: (1) it may use functional terms implicitly presupposing higher-level truths or (2) it may directly use higher-level information in its explanations.[2] For example, an individualist theory of society that employs social role predicates—such as "teacher" or "policeman"—would face the first problem, since these terms arguably presuppose sociological facts about institutions. In this case, predicates could refer to individuals and yet still not suffice for reduction—not simply because they appeal to functions but because they invoke functions in higher-level systems described in higher-level terms. In short, theories that describe roles or functions fail to reduce if those functions describe roles in systems described in higher-level terms. Lower-level theories may also directly invoke higher-level facts. A neurological theory, for example, that used information about conscious awareness, motivation, and goals would be using information about psychological states. Once again, reduction is thwarted, for the higher-level theory is presupposed rather than replace with a purely lower-level account.

Each of the above problems must be empirically established and may not be by itself decisive, since there are conceivable routes around them. In the following sections I will try to show that those problems are in fact real and unavoidable in molecular biology. Before turning to that task, it is necessary to answer one more preliminary question: what shall we take the reduced and reducing theory to be? Molecular biology has traditionally been taken to be the reducing theory, but that seems to me to be a misnomer. Molecular biology is not simply biochemistry; it constantly and essentially refers to biological entities and biological functions. Organelles and other biological entities as well as their biological functions are part and parcel of standard molecular biological explanations. (By "molecular biology" I mean that body of theory taught in the standard cell biology class and summarized in textbooks such as the *Molecular Biology of the Cell* [Alberts et al. 1994]; I do not intend the more narrow meaning, common among some cell biologists, which equates molecular biology primarily with the mechanics of DNA expression.) Molecular biology thus seems to be the candidate theory for reduction, with biochemistry roughly being the reducing theory.[3]

The Signal Hypothesis

Although molecular biologists can confidently describe how DNA transcription gives rise to polypeptides, only recently have they begun to say how the protein "knows where to go" in the cell. Proteins are synthesized in the cytoplasm by complexes of ribosomes and mRNA. The protein must then be transported to its ultimate destination. Proteins destined for secretion are translocated across the endoplasmic reticulum (ER) membrane and eventually reach the plasma membrane; other proteins are incorporated either into the ER membrane or transported to mitochondria, chloroplasts, or elsewhere. In each case the protein must go from its location in the cytoplasm to its final destination. Somehow the protein must encode the information that determines where it goes.

The work of Blobel and Dobberstein (1975) first showed how that information is stored and expressed. Working with a cell-free protein synthesis system, they found that immunoglobulin is initially synthesized as a larger precursor. The precursor contains a sequence of amino acids that is synthesized first. Upon emerging, it directs the ribosome complex to the ER membrane and meditates translocation of the new protein across the membrane and into the lumen of the ER. This initial sequence thus serves as a signal for sorting proteins.

Blobel and Dobberstein thus proposed the "signal hypothesis": the information determining protein transport is contained in a signal sequence, a particular sequence of amino acids incorporated in the protein precursor itself. Since the initial work, signal sequences have also been found that mediate protein transport to chloroplasts, mitochondria, and nuclei in a variety of cell types and species (Sabatini et al. 1982; Goldfarb et al. 1986). Details of the model have also been refined. For proteins that are not transported into the ER lumen but inserted into the ER membrane, a second signal has been found (the stop transfer sequence) that stops transfer across the membrane after insertion (Friedlander and Blobel 1985). A signal recognition particle as well as receptor proteins on the ER membrane are also involved in binding the ribosome to the ER (Walter et al. 1981; Ng and Walter 1994). The signal hypothesis continues to pick up further confirmation and refinement as a major mechanism of protein targeting.

How does the signal hypothesis bear on reducibility? Each of the possible obstacles facing reduction listed above are real ones for any attempt at a purely biochemical account of protein targeting: (1) "Signal sequence" is a predicate that has multiple realizations in biochemical terms. At least two hundred different amino acid sequences have been found to act as signals. Searches for common motifs in those sequences have failed (see, for example, Gal and Raikel 1993). (2) Whether a particular signal sequence actually functions as such depends on the cellular context. Sequences that serve as a signal for one protein are also found in other proteins without playing that role. (3) "Signal sequence" is defined in terms of its biological function in molecular biology: it is the set of amino acids that determines to which organelle or other cell location the protein is transported. This definition, however, is not purely biochemical, for it essentially refers to organelles and other biological components. Thus, explanations employing "signal sequence" cannot be employed unaltered to reduce molecular biology, even though the term refers to some set of amino acids. Furthermore, the other components of the signal hypothesis—"stop transfer sequences," "signal recognition particle," and "receptor proteins"—may also exhibit the same difficulties, particularly problems (1) and (2).

However, we have already seen that each of these problems is strong prima facie evidence against reducibility. Because "signal sequence" is defined via its biological function, biochemical explanations that employ this predicate will presuppose biological facts. Reduction will thus be defeated. To avoid this difficulty, the reductionist must supply biochemical explanations in a much more austere language—and yet still be able to capture the biological function of a signal sequence in purely physical terms. Doing that, of course, runs up against the obstacles of multiple realizations and content sensitivity. Unless multiple realizations can be classified or otherwise delineated, they prevent the reductionist from equating biological terms with biochemical counterparts. Even if counterparts can be found, context sensitivity means that biochemical descriptions will not provide the appropriate reductive definitions.

There are, of course, various general strategies for circumventing these obstacles to reduction. The empirical evidence, however, strongly indicates that all are dead ends. Let me discuss them in turn.

One possible way to avoid the multiple realizations problem would be simply to define "signal sequence" in terms of the known amino acid chains that serve this function. We could, it seems, avoid the multiple realizations problem and eliminate reference to biological functions by equating "signal sequence" with the known sequences serving the relevant role. Hull (1974), for example, apparently thinks some such move makes multiple realizations unproblematic.

We saw in the last chapter that disjunctive definitions of this sort are inadequate, because they do not capture the counterfactual force of laws and generalizations. "Proteins are sorted by a signal sequence" makes a claim not just about proteins studied so far but about future, as of yet unobserved instances. Equating "signal sequence" with its known realizations would not adequately preserve the relevant generalizations. A disjunctive definition would be adequate only if we know all *possible* signal sequences. But that seems unlikely, since the number of signal sequences is quite large and there is no known way to delimit what future discoveries will be. We thus cannot produce an adequate disjunctive definition.

A different route around the multiple realizations problem would be to grant that there is no unique definition of signal sequence in biochemical terms but deny that this fact causes problems. Multiple realizations, so the argument might run, only show that the biological concept of signal is inadequate. We need to replace it with a much more refined conceptual scheme that recognizes many signals—one for each underlying biochemical mechanism. Once this change is accomplished, the multiple realizations problem disappears, for there is now a unique biochemical property for our newly revised signals.

Nothing rules out such a strategy a priori. But the empirical details make it highly suspect for the following reasons: (1) While subdividing biological concepts may sometimes be fruitful, the most obvious and natural way to fragment "signal" would be by distinguishing secretion signals, mitochondrial targeting signals, nuclear localization signals, and so on. However, such a subdivision is done on *functional* lines, not biochemical, and as such does not by itself give us a biochemical definition. (2) Even within these subclasses of signals, there is apparently little prospect of refining away the multiple realizations problem. The reason is that the signal function—even if we recognize a diversity of signals—is brought about by too diverse a set of biochemical properties. The evidence here is complex and extensive, but consider the following four facts:

1. Signal sequences for insertion into the ER can be either co- or posttranslational, sometimes do and sometime do not depend on features of the protein itself to serve their function (Lehnhardt et al. 1987); may vary over 200 percent in length; apparently show diverse physical chemical interaction with membrane lipids (Batenburg et al. 1988; Fidelio et al. 1986); may or may not be cleaved in serving their function depending on the signal sequence involved; may or may not require the presence of a signal recognition particle; may or may not require the presence of molecular "chaperones" (Hohfield and Hartl 1994); and are sometimes species specific in their functioning (Powell et al. 1988; Oliver 1985).
2. Mitochondrial targeting signals show a similar diversity (see Hay et al. 1984 and references cited therein): such signals are species-specific; involve different mechanisms for different proteins; may lead to insertion into either the inner membrane area, the outer membrane, the inner membrane, or the mitochondrial matrix; may or may not suffice to determine the internal mitochondrial destination; invoke different pathways depending on ultimate destination; and sometimes require electric potential and sometimes do not.
3. Signal sequences may either be leader sequences or internal to the protein itself; internal and external sequences are apparently structurally disparate (Dalbey and Wickner 1987). Signal sequences may also occur at either the carboxyl or amino termini of the protein (Chrispeels and Raikel 1992).
4. It is not at all clear that defining "same biochemical mechanism"

can be done in purely biochemical terms—something we would have to do if we wanted to avoid multiple realizations by subdividing signals according to underlying physical structure or conformation. Most mechanisms involving signal sequences will depend on weak, noncovalent bonds. But such bonding need not be an all-or-nothing process, since it may result from the combination of multiple binding sites or be nonadditive (see Creighton 1984). Consequently, it may well be that "same mechanism" or "same physical conformation" will be defined not in purely biochemical terms but rather in terms of whether the biochemical structure in question binds well enough to serve the signal function. So in specifying the same mechanism or physical structure we would be reintroducing, rather than eliminating, biological information. Since there is evidence that different sequences serve the signal function with varying degrees of effectiveness (for example, Kaiser et al. [1987] found that for some proteins 20 percent of *random* human DNA sequences will to some extent successfully encode a signal sequence), this problem seems likely.

What do these facts tell us? Problems (1)–(3) show that there is much biochemical diversity even after we distinguish various kinds of signals—that the multiple realization problem reappears even if we recognize different sorts of signals. Such diversity makes any further attempt at fragmentation of "signal" a hopeless project. We would have to subdivide in great detail, recognizing signals for each kind of mechanism and submechanism, destination, and species. Cell biology would be able to say little of a general nature even about such subdivided signals as "mitochondrial signals"; it would be reduced to simply listing known biochemical facts. If the biological supervenes on the physical, then such a list is always possible. But it has little to do with reduction, for we have not captured the generalizations, categories, and explanations of biology, we have simply abandoned them. And problem (4) shows that even if we could avoid such a radical fragmentation, definitions in terms of mechanisms may implicitly presuppose biological information.[4]

Attempts to avoid the context-sensitivity problem run into similar difficulties. It may seem obvious that even if one and the same sequence of amino acids can sometimes serve as a signal and sometimes not, such differences in context must have a biochemical explanation. That may well be, but noting this fact does not necessarily remove the context sensitivity problem for two reasons: (1) Further specifying the context may explain the difference, but we must be able to specify the context without invoking biological information or else we have not given a fully biochemical explanation. However, it may often be that the relevant context does indeed invoke biological information—as it would if we explained any context sensitivity in part by invoking the fact that the protein was bound to a receptor or membrane (the latter terms are from cell biology, not biochemistry). (2) Even if we avoid explaining contextual effects directly in biological terms, we may well do so indirectly. To explain contextual effects by saying that the protein in question had a different conformation requires us to identify which conformations count as signals and which do not. As I argued above, having the "right" conformation for being a signal sequence may be defined in terms of a sequence's ability to serve the relevant biological function rather than in purely biochemical terms.

Thus there is evidence that the context sensitivity problem cannot be easily avoided, although a more decisive evaluation awaits further empirical details. Note, however, that even if we can in each case explain the contextual effect in biochemical terms, the problems of multiple realizations and presupposing biological information remain. For having a purely biochemical account of each particular signal sequence is completely compatible with there being no identifiable defining biochemical property for the predicate "is a signal sequence" or its subdivisions; the evidence given above strongly indicates that this is the case.

Finally, let us consider a final route around the problem of presupposing biological information: if "signal" is defined in terms of its biological function, then we need only, in turn, to explain those functions in biochemical terms. Of course, nothing automatically rules out that prospect. But the evidence on the other side should be clear. We would need to define in biochemical terms all the biological information used in explaining the signal hypotheses. However, that information is extensive: it involves multiple organelles, membranes, functional pathways, receptors, and stop transfer signals. Redefining this biological information—much of which is again depicted in functional terms—will also potentially face all the problems cited so far: multiple realizations, context sensitivity, and implicitly presupposing biological information. Those reductions could of course run smoothly, unlike the signal hypothesis. The next section, however, will go some way toward showing that prospect to be unlikely.

Thus it seems that the problems for reducing the signal hypothesis remain in force. It is an empirical fact that they do so, but an empirical fact that seems to make good evolutionary sense. Given that natural selection can often be blind to underlying mechanisms and allow diverse structures so long as they produce the same effect, we would expect that similar functions might be brought about by sundry biochemical mechanisms and that it is the functional similarities that diverse species, cell types, or organelles have in common, not the biochemical details (see Rosenberg 1989). But such biochemical diversity underlying biological unity is the root obstacle to reduction.

Irreducibility Is Rampant

The signal hypothesis presents a clear example of the obstacles to reduction. Nonetheless, the problems it illustrates are not unique—they pervade molecular biology. In this section, I support this assertion by describing a number of results in molecular biology: work on cell communication, the immune system, embryonic development, and cell structure. Each area exemplifies the obstacles to reduction found in the signal hypothesis.

Cell Communication

Cells receive information from their external environment and respond appropriately (see Snyder 1985; Berridge 1985; Alberts et al. 1994, chap. 15). Molecular biology has made important strides in understanding how cells receive and process information. In the broadest outlines, external signals are brought to the plasma membrane where they are recognized by receptor proteins on the cell surface. The relevant information is then communicated, usually via a mediating protein, to a second messenger that serves to carry the signal from the cell surface to its intracellular destination.

As may be obvious, this account describes the cellular and/or organelle-level functions of molecular components. Signals are molecules that pass *information* to the cell. Receptor proteins function to capture such signals and pass information from the receptor across the cell *membrane* to the second *messenger*. Second messengers transfer the signal to the appropriate *organelle*. In each case these explanations presuppose biological truths. As a result, any biochemical theory employing them would not suffice for reduction.

Eliminating these partially higher-level explanations seems unlikely, for no obvious biochemical definition is forthcoming: each functional predicate has multiple realizations. Signals may be small peptides, proteins, glycoproteins, amino acids, steroids, or fatty acid derivatives (and in the case of neurosignals, at least fifty have been identified so far). Furthermore, not every chemical falling into these classes serves as a signal and those that do carry different information depending on the cellular context (so any attempt to define a particular signal also faces problems of context sensitivity). Receptor proteins must be specific to signals and thus likewise seem to have indefinitely many realizations. Not every receptor is a protein, since glycolipids also apparently serve the same functions. Finally, a great many different proteins function as transport proteins depending on cell type and other such biological factors.

Immune System

The heart and soul of the immune system is the antibody that binds to the antigen—a foreign body such as a bacterium—and thus facilitates removal (see Alberts et al. 1994; Tonegawa 1985). Antibodies are receptor proteins attached to the surface of lymphocytes; they serve to bind the antigen to the lymphocyte, stimulating the production of new lymphocytes and also functioning as markers for macrophages. "Antibody" is obviously defined in terms of biological functions and entities (which probably reside at multiple biological levels, for example, of the cell, organ, and organism). Consequently, any biochemical explanation employing this term will not suffice for reduction.

Can this obstacle to reduction be eliminated by defining antibody biochemically? It seems unlikely. Antibodies have potentially millions of different chemical structures—estimates are that humans make 10^{15} different antibodies. The multiple realizations problem is very real indeed.

The diverse chemical structures serving as antibodies are, at least in higher vertebrates, all modifications of a few underlying forms. Does this partial isomorphism suffice to physically define "antibody"? It does not. The general physical structure common to all antibodies—roughly a Y configuration—is not sufficient to define "antibody," even if we can group antibodies according to their approximate physical structure. In the first place, the Y structure is not universal, for antibodies can cross-link to form more complex structures. Moreover, it is not this general structure that makes a protein an antibody. Rather, the protein has a certain region or sequence of amino acids that allows it to recognize a specific antigen. Each of these regions has a unique physical structure; they may bind one or many different sites on the antigen (nearly any macro molecule can be an antigen, so the possible antigenic structures are equally open ended). There are as many of these different physical structures that bind as there are kinds of antibodies. So the multiple realizations problem remains.

Furthermore, even if we had some physical way of defining all those structures that bind antigens, we would still not be out of the woods. Merely binding to an antigen does not constitute an antibody. Antibodies must not only bind but also elicit an immune response—there are substances that bind with no effect. Binding results from weak noncovalent bonds as in the signal case, and thus is a matter of degree. Binding itself does not cause an immunological response; it is rather the interaction of the other parts of the antibody with further elements of the immune system. The conclusion is thus inescapable: "antibody" must be defined via its biological function, not physically.

Cell Structure

The conformation of the cell is determined in large part by the cytoskeleton, which apparently also serves nonstructural functions (see Alberts et al. 1994; Weber and Osborn 1985). The cytoskeleton consists of three parts: actin filaments, microtubules, and intermediate filaments. Again, this minimal account raises the problems of functional terms and multiple realizations. Actin filaments, for example, require actin-binding proteins—proteins that serve to give the cytoskeleton the proper structure. Actin-binding proteins are thus defined via their biological role. The proteins that play this role are in part specific to each cell type, thus raising problems of multiple realizations. While actin-binding proteins can be subdivided, those subdivisions (stabilizing factor, spacing factor, etc.) refer to their cellular function, not their biochemistry.

Embryonic Development

Explaining embryonic development raises the standard problems for reduction (see Alberts et al. 1994; Gehring 1985; Fjose et al. 1985). While the details of development are still largely unknown, some initial explanations are available for lower organisms. For example, embryonic development of the fruit fly has established that there are specific genes controlling segmentation patterns and the ultimate functions of larval units (names segmentation and homeotic genes respectively). Segmentation and homeotic genes can be grouped under a more general class of regulatory genes—genes that regulate cell functions by switching off and on other genes. As may be obvious, such genes are identified by their cellular function, and those functions are often biologically specified. Homeotic genes, for example, are those that determine for a specific imaginal disc what *organ* it will ultimately become, what its ultimate location in the *organism* shall be. Homeotic genes in turn require information about their location (that is, the location of the nucleus carrying them) vis-à-vis other nuclei. Such information is apparently provided by a "sensor"—an upstream DNA segment that is turned off and on by information about cell location. Thus, even if we had a biochemical equivalent of "sensor," explanations of how sensors worked would use biological facts about cell location.

These regulatory genes controlling development are obviously specified by their biological function. If there are general processes controlling development across species, then we can expect such functional specifications—rather than biochemical definitions—to be the rule, for the precise molecular details may vary as each species' pattern of development varies. Homeotic genes, for example, differ from 20 to 40 percent in their sequences from one species to another.

Of course, none of the above examples shows molecular biology to be forever irreducible. Further advances in biochemistry may allow us to group multiply realized predicates in ways currently unavailable; functional terms could conceivably be replaced. But these possibilities are merely that. The best evidence is that molecular biology is here to stay for reasons that go far beyond pragmatic difficulties. Molecular biology groups together—via terms describing biological functions—cellular phenomena that are physically quite diverse. Physically similar phenomena differ in their biological import, because the biological content differs. Molecular biology and its likely permutations appear irreducible.

Nothing in these conclusions entails, however, that the molecular mechanisms underlying cellular processes cannot be studied and explained in detail. Molecular biology in part seeks such explanations. But explaining molecular realizations of biological processes in no way entails that the latter are reducible to the former, as the previous discussion amply illustrates. Molecular biology provides molecular details without supporting reduction. It is this bilevel nature of molecular biology that makes it a particularly good model for nonreductive scientific unity, or so I shall argue later.

Before we move on, I want to consider one last objection to the arguments of this section. Someone might grant my claims that molecular biology is not reducible but argue that this just shows that molecular biology is scientifically inadequate. We have seen variants of this strategy before when we considered a generalized eliminativism about social theory. Rosenberg (1994) has argued for some such position about biology, and I need to consider his arguments, for they threaten to undermine the significance of my results.

Rosenberg argues that because the biological only supervenes on the physical, it is bound to be scientifically limited. The diverse sorts of physical realizations that I argued showed nonreducibility Rosenberg thinks show something more—namely, that the biological sciences will never produce strict laws and are not indefinitely refinable in the way that physics is. Thus, he argues, we should conclude that the biological sciences are largely only useful instruments, not explanatory theories. Many others share Rosenberg's intuition that the high-level sciences just don't have the right relation to the physical to be full-fledged explanatory enterprises (see Searle 1981).

These intuitions are unconvincing for several reasons. First, they depend on a picture of the physical sciences that I find questionable. As Cartwright (1984) has argued in detail, ceteris paribus qualifications abound in our best physics. If its basic laws are taken strictly, then they are false; they explain real processes only by adding piecemeal qualifications needed for the particular case at hand. Thus, the idea that physics produces strict laws is suspect and thus is no reason to doubt the scientific standing of the biological sciences.

Of course, the conclusion to draw here is not that the physical sciences should likewise be treated instrumentally—as nonexplanatory—but that we have gone wrong somewhere in our analysis. The error lies in assuming that nonstrict laws or generalizations mean inferior explanations. That is not the case. The clearest and most central form of explanation comes from citing causes. Yet we have abundant evidence from everyday life and from the practice of all the sciences that causes can be identified without having exceptionless laws of universal scope. I can confidently assert that my flying foot caused the broken door even though I have no idea how to state that as an exceptionless law, and I can identify smoking as a cause of cancer even though I have no strict law relating cancer and all its prior causal factors.

Moreover, supervenience without reduction does not rule out predictive improvement. The physics of macrosized objects has no reduction in quantum mechanical terms, and as we saw in the first chapter thermodynamics is apparently not reducible to statistical mechanics. Still there has been enormous improvements in predictive power in both these areas. There has likewise been such improvement in molecular biology, despite the fact that many of its kinds have multiple realizations in physical terms. Improvements in immunology where the main kinds or categories clearly have indefinitely many physical realizations is a clear example. So the nonreductive nature of the biological sciences does little to show that we should view them instrumentally.

Reductionist Heuristics

So far I have argued against reducibility considered as an explanatory relation between theories, with the latter being considered sets of statements. This section looks at other elements of molecular biology, in particular, heuristics for discovery and methods of justification. The arguments given so far strongly suggest that purely biochemical approaches to cell behavior are sometimes heuristically ill advised and are lacking important methods of confirmation.[5] Of course, few philosophers, even those who defend reduction, argue that biological *research* should be carried on using only lower-level information. Nonetheless, reductive research strategies still find significant sympathy among practicing biologists, especially molecular biologists. And reductive strategies are quite commonly advocated in the social and behavioral sciences, sometimes on the grounds that biology proceeds reductively (see Elster 1986). So it is worth explaining how the failure of reductive explanations bears on scientific practice.

Purely biochemical approaches can often be heuristically unsound: they involve implicit approaches to investigation that are quite likely to be unfruitful. While irreducibility does not logically entail that purely lower-level approaches will lead investigators astray, it does argue against their general fruitfulness. There are at least three reasons to think that biochemical theory will obstruct discovery.

Attempts to explain in purely biochemical terms will tend to see diversity where there is important unity. Because the kinds employed by molecular biology often have multiple realizations, proceeding biochemically will not reveal common mechanisms and explanations. Rather, the biochemical data will indicate that the processes under investigation differ in kind.

For example, recent work on the signal hypothesis led to the discovery of signal sequences internal to the transported proteins (Freidlander and Blobel 1985). Even though the complete sequence of opsin was known, it was impossible to distinguish between signal sequences and transmembrane sequences from the biochemical information alone; this fact was only determined from a biological assay. By using a biological assay that copies cell function in the test tube, internal segments of opsin were found to function as signal sequences. It was looking for the *biological function,* not the biochemistry alone, that allowed this discovery. Because signal sequences are biochemically diverse in nature, studying just the biochemistry of opsin would have been a dead end. Unity lies at the biological level.

Biochemical theories will also face problems on the other side of the coin: they will see unity where there is none. For example, if we try to understand protein targeting by concentrating solely on the physical chemistry of signal sequences, then we will naturally be led to investigate other proteins containing those sequences. But signal sequences do not always function as such. We will accumulate information irrelevant to the task at hand—explaining protein targeting.

Finally, biochemical theory is heuristically unsound because it lacks what might be called biological "place holders." A "place holder" is a term of a theory that is introduced to designate an entity or process for which (1) we have good evidence but (2) we do not know its precise nature. For example, "transforming principle" was an important place holder in biological theory before the role of DNA was discovered. Often biologists have evidence that some biological function is carried out long before they know the details. Progress often comes from describing such functions, designating a place holder, and then investigating it in biological and chemical terms.

A purely biochemical approach would have us give up *biological* place holders—place holders that identified cellular rather than biochemical roles. In so doing, an important heuristic tool would be lost. As we have seen, molecular biology is rife with terms describing biological functions, and many important results have come from postulating a biological process and then investigating it. The signal hypothesis is a prime example in this regard. Insofar as we search for purely biochemical explanations, we forgo this important heuristic tool.

Purely biochemical theories will face confirmation problems from at least two directions. Given that important molecular biological kinds will have multiple realizations, then biochemical theories will naturally tend to produce unwarranted generalizations. To explain the general process of protein targeting, biochemistry would have to make recourse to disjunctive statements such as "proteins are targeted by one of the amino acid sequences S_1 or S_2 or . . . S_n". However, such disjunctive generalizations will of course be disconfirmed every time a new instance of a signal sequence is found.

More important, the problems for reduction discussed in previous sections show that eschewing all biological theory will deprive many biochemical explanations of their confirming evidence. Experiments and data confirm a statement only when conjoined with background theory specifying the relevant factors of experimental design—the relevant variables and factors that must be controlled. However, much of what we know about the biochemistry of cells relies heavily on molecular biological background theory to specify experimental design. In short, biochemical theory often must employ molecular biological theory to design its experiments and thus to achieve confirmation.

Work on the signal hypothesis again provides a clear example. Take a biochemical explanation of protein targeting; the cellular destination of light chain immunoglobulin is determined by a twenty amino acid sequence at the amino terminal end of the translated precursor.[6] How is this statement confirmed? It was not purely biochemical information that did the trick but rather a biological assay that assumed biological information about the endoplasmic reticulum, ribosomes, and the kind of cells involved. Confirming this simple biochemical explanation requires a great deal of molecular biological theory.

It should not be surprising that molecular biology plays this role. We often explain the functions of molecules by appealing to their role in the *cell* and other biological entities: we seek biochemical explanations of biological functions that have no unique or invariant molecular equivalent. As a result, we study the biochemistry of the cell by employing what we know about its biology, without being able to eliminate the latter in the process of confirmation. Biochemistry needs molecular biology.[7]

Nonreductive Unity

While I have denied that molecular biology is advancing the unity of science by reducing biology to chemistry, molecular biology is nonetheless advancing the cause of scientific unity. Obviously, that is possible only if there can be unity without reduction. Molecular biology, I shall argue, provides an excellent example of such nonreductive unity. That unity basically involves interconnections and dependencies between theories that nonetheless do not allow for one theory to replace the other. Those interconnections can occur in at least the following seven ways (listed roughly in increasing order of significance):

1. The ontology of one theory may exhaust that of the other—every entity described by the one theory may be composed of or token identical to some entity described by the other.
2. Two theories may be irreducible yet logically compatible in that neither explicitly or implicitly uses or presupposes statements at odds with those from the other theory.[8]
3. One theory may supervene upon another: the referents of the basic predicates and the truths of one theory fix or determine those of the other.
4. It may not just be the case *that* supervenience holds but that we can say *how* it does so—in other words, we can cite some of the mechanisms that realize or bring about higher-level kinds (thus providing inductive evidence for claims of supervenience).
5. One theory may be heuristically dependent upon another—it may use the other theory to suggest fruitful avenues of research.
6. One theory may be confirmationally dependent upon the other—it may use the other theory in experimental design.
7. One theory may use, explicitly or implicitly, explanations from the other. Each of these conditions describes a sense in which a theory is not autonomous; all are logically independent of reducibility.

Unity reaches its pinnacle when (1)–(3) obtain and when (4)–(7) hold between both theories, that is, in both directions, and do so extensively. When these requirements are met, then the two theories are linked by, or perhaps incorporated in, an *integrated interlevel* theory.[9] An interlevel theory thus unifies two disparate theories by employing explanations and confirmation procedures invoking both levels and by providing evidence that the events and entities of one theory depend upon and are constituted from those of the other.

Note that these connections may hold completely between two theories and yet neither theory may be reducible to the other. If we stick to the root notion of reduction—that one theory can do all the work of or replace another—then unity as just described in no way entails reduction. For one theory to do all the work of another, there must be some systematic way of replacing higher-level explanations with lower-level explanations. But doing that requires the appropriate bridge laws. None of the above criteria, either singly or jointly, entails that such laws can be provided. As the examples of the previous sections clearly illustrate, two theories can be quite interdependent while reduction seems impossible.

Molecular biology provides a prime example of an integrated interlevel theory. It unites biological theory at the cellular level with biochemistry while leaving the biological aspects of its explanations unreduced.

Molecular biology has obviously dealt the finishing blows to vitalism. It has provided overwhelming evidence that every living thing is composed of chemical constituents. Likewise, detailed study of the molecular processes underlying the biological activities of the cell give us good reason to think that biological facts supervene on those of biochemistry. Tracing out these dependencies gives us what Kitcher (1984) has called "explanatory extensions." We can see how cellular processes are possible; we come to answer an important range of questions—how-questions—that purely biochemical accounts are wont to answer.

We have already seen that biochemistry depends upon biological information about cellular behavior. As was argued in earlier sections, biochemical accounts often invoke biological truths in their explanations. They also employ information from the cellular level both as a heuristic to suggest new avenues of research and in the construction of experimental design. Thus, biochemistry depends upon the biological theory of the cell in ways (5)–(7) above.

The dependence, of course, runs the other direction as well. Biological assays require all kinds of biochemical information to confirm their results. Explanations of cellular processes in molecular biology constantly invoke biochemical information, and postulating possible biochemical mechanisms often suggests fruitful avenues of research into cellular-level processes.

So, while molecular biology does not support the reducibility of biology to chemistry, it certainly argues for their unity. Molecular biology leaves biological explanations their role, yet combines them in a systematic fashion with those of biochemistry. As such, molecular biology presents a paradigm of nonreductive unity in the sciences.

5

Defending Nonreductive Unity

So far I have defended the idea that the special sciences are not generally reducible to their lower-level counterparts. Nonetheless, I have claimed that this irreducibility need not commit us to vital forces or to social entities acting behind the back of individuals. Even though the special sciences are irreducible, they are dependent: high-level entities like classes are composed from their lower-level counterparts and supervene upon them. This dependence, moreover, leaves lower-level theories an important place in explanation even if the special sciences are irreducible.

To many this story looks too good to be true. Critics doubt that these combinations of claims are compatible. Some argue that dependence—in particular, supervenience—entails reducibility. Some argue that autonomous special sciences commit us either to an incoherent notion of downward causation or leave higher-level properties causally impotent. Other critics argue that even if reducibility fails, supervenience ensures that lower-level theories can fully explain; a full account of the supervenience base is a full explanation. Finally, some antireductionists argue that supervenience leaves lower-level accounts no explanatory role, thus likewise challenging my happy story that gives lower-level detail a place.

This chapter rebuts these charges. First I examine the arguments for the claim that supervenience entails reducibility. I argue that once again critics confuse ontological claims for epistemological ones and in an important sense they also beg the question. Next I take up various arguments purporting to show that autonomous special sciences entail metaphysical absurdities—for example, that there are no causes above the physical level and that the special sciences cannot describe natural kinds. Following this I set up the issues in the debate between those who claim that supervenience ensures full explanation—regardless of reducibility—and those who claim it allows for no explanation. I then argue that explanations in lower-level terms are drastically incomplete, for they cannot capture important causal patterns. I finish by answering those extremists who think lower-level accounts explain nothing.

Does Supervenience Entail Reducibility?

Many philosophers (e.g., Rosenberg 1985; Bacon 1986; Kim 1993; Grimes 1995) have tried to show that supervenience actually entails reducibility and thus cannot provide a nonreductive picture of scientific unity. The crux of their argument is this: If the facts about *W* supervene on the facts about *P*, then there is a lawlike connection from *P* to *W*. That lawlike connection can be turned into the bridge laws needed for reduction, even if *W* facts are multiply realized in *P* facts. If we take the disjunction of all the different supervenience bases realizing *W*, then we will have our bridge laws and thus reduction. So we won't have a simple biconditional of the form:

$$W \text{ if and only if } P$$

But we will have a complex biconditional of the form:

$$W \text{ if and only if } P_1 \text{ or } P_2 \text{ or } P_3 \text{ or} \ldots$$

But we have no reason to think reductions should be as simple as (1); complex connections like (2) will still do the job. Supervenience entails reducibility and my version of nonreductive unity is incoherent.

This argument turns on confusing ontological and epistemological claims; it also ignores the obstacles that face real reductions in the sciences. In the process it also begs the question. I work through these problems in turn:

1. Bridge laws connecting a higher-level kind with its known realizations will not do. Bridge laws must have modal or counterfactual force—they must tell us not just what has happened but what would or will happen. If I wanted, for example, to reduce explanations in terms of fitness to purely physical explanations, it would not do to equate fitness with all the various physical traits realizing it so far. My explanations involving fitness make claims about

what will happen in the future—the most fit future individuals will survive over the less fit—and they make claims about what would happen: if organism *X* had the same fitness but a different physical trait, it would have survived. Neither of these claims turn out true if I equate fitness with its actual realizations so far. The ontological facts about how *X* has been realized so far do not capture the explanatory or epistemological claims we want to make about fitness.

2. Thus bridge laws must connect higher-level kinds with at least their possible physical realizations. This raises several problems. For some higher-level predicates, it may well be that the number of possible physical realizations is infinite. Infinite disjunctions seem odd candidates for a reducing predicate. At issue here is more than the fact that disjunctive definitions are likely to be complex and unusable for practical purposes, something reductionists will grant. Rather, the problem lies in the fact that commitment to the metaphysical principle of supervenience does not justify the epistemological claim of reduction. Reducibility in principle requires, I take it, at least that someone or some group of individuals could produce the relevant definitions, if provided with sufficient computational and memory resources. So imagine a reductively minded individual with the computational and memory abilities of God. Give this divine reductionist just the fact that higher-level theories supervenes upon finitely many lower-level realizations. Would he or she be able to provide a disjunctive definition for each higher-level predicate? The answer seems to be a clear no. Even granted infinite computational ability, knowing the principle of supervenience does not give us a function to generate all physically possible supervenience bases. In short, knowing that *W* supervenes on *P* does not entail I know how it does so. Reducibility is obstructed, not by the complexity of the definition, but by the fact that supervenience by itself gives us no systematic way, even in principle, of supplying all the relevant disjuncts. And this problem holds even when we grant that the number of supervenience bases is finite.
3. Even if we could list all possible supervenience bases, reduction would not be ensured. That is because reductions require not only lawlike bridge laws but also require bridge laws that allow lower-level theories to do all the explanatory work of higher-level theories. As such, supervenience may still give us not suitable bridge laws. I can see at least three such possible situations: (1) where

> the supervenience base in effect allows too much in, that is, when it fixes or determines more than one higher-level predicate, (2) where the lower-level predicates in question individuate in ways very different from that of the higher-level predicates, and (3) where the predicates of the reducing theory presuppose predicates or truths from the theory to be reduced.

A supervenience base may allow "too much in" in the following sense. A property or fact *W* supervenes on some other property or fact *P* just in case fixing *P* also fixes *W*. However, nothing in this relation rules out the possibility that other properties or facts may also supervene on *P*—that *P* may also fix or determine other facts or properties. Once we see that multiple properties can supervene on the same base, then we can also see that definitions in terms of realizations may not support in principle reducibility. For example, imagine that we have the following truths:

1. Events of kind W_1 cause events of type *E*.
2. Events of kind W_2 do not suffice to cause events of type *E*.

Imagine also that in some particular instance W_1 and W_2 are realized on the same supervenience base, *P*. Now if we define W_1 and W_2 in terms of their physical realization, our reduced theory will entail that *P*— and thus W_2—causes on some occasions events of type *E*. Consequently, our reduction will not preserve the relevant truth values as it must.

Consider the following commonsense example of the problem just sketched. Two properties of water—its capacity to conduct electricity and its translucence—supervene on its molecular structure. Imagine an electrical short caused by the presence of water. If we equate both properties with the description of their supervenience base, that is, with "is H_2O," then we end up with the conclusion that the water's being translucent caused the electrical short. Of course, that is absurd—the water's translucence is causally irrelevant. In short, identifying supervening predicates with their bases for the purposes of reduction may lead—when one base determines multiple properties—to collapsing properties that are causally distinct.

Note that the problems caused by one supervenience base determining multiple supervening states remain even if those two states have differing supervenience bases in some circumstances. Translucence, for example, is not always realized in the structure of water nor is the ability to conduct electrical current. Yet those properties or predicates do in some instances supervene on the same physical structure. That means any attempt to equate them with their supervenience bases will produce false results. If I define conductivity by "having the molecular structure of H_2O or the structure of Hg or etc." and I define translucence by its supervenience bases ("having the molecular structure of H_2O or O_2 or etc."), then any causal generalization about conductivity will entail that translucence sometimes has that causal effect (since conductivity equals having the structure of H_2O equals being translucent) even though conductivity and translucence do not have coextensional supervenience bases in all possible contexts. In short, this problem does not require that the two higher-level predicates have either all the same physically possible or logically possible realizations.

So, while there may be further stipulations that would rule out the kind of problem presented by one supervenience base determining multiple properties, that would only show that mere supervenience does not ensure the appropriate reductive definitions. Supervenience of its own allows one physical base to fix multiple biological properties, and that prospect means reductions formed from disjoining supervenience bases may fail because they allow in the wrong information. To make this point in another way, advocates of reduction via disjunction forget that reducibility requires not just bridge laws between predicates but bridge laws that allow the reduced theory to be derived from the reducing theory. However, as the examples cited above indicate, definitions in terms of supervenience bases may not provide for the appropriate derivations.

A second problem arises from the way these arguments formulate the notion of supervenience: a property *W* of *X* supervenes on some property or properties *P* just in case whenever *X* has *P* it also has *W*. Putting supervenience this way implicitly makes an important assumption, namely, that the extension of the predicates of the high-level theory apply to the same objects as do those of the reducing theory. Why does this assumption matter? Higher-level facts might well be supervenient upon the physical facts even though in every particular case the two kinds of predicates cut up the world very differently or, in other words, have extensions that only overlap. As a matter of logic, nothing rules out the possibility that the significant units for higher-level theories such as biology or sociology may never be precisely equivalent to those of physics, for example. If the latter situation obtains, then it becomes difficult to define higher-level terms by their lower-level realizations, for our definitions will again include "too much" (or too little).

The relation between Mendelian genes and strands of DNA provides a likely example of this problem. Mendelian genes presumably supervene upon DNA: two organisms with the same DNA would also have the same genes. Nonetheless, the relation between genes (conceived as directly correlated with phenotypic traits) and DNA is enormously complex. The DNA strands connected with some gross phenotypic trait will not be simply one continuous sequence of bases. Rather, a number of different sequences producing different proteins with multiple functions along with various nontranslating sequences such as promoter regions will be responsible for the phenotypic trait. Those component sequences of this complex may well have other roles and be involved in other complexes to produce other phenotypic traits. Given this complex mess, there may be no obvious set of sequences to call "the" gene, even if all the facts about DNA sequences determine what genes are had.

The root of the problem here is that supervenience is being asked to do more work than it legitimately can. Supervenience on its own only requires that one kind of fact (or property, case, etc.) fix another; it says nothing concerning what those facts are about—they logically could be about entirely disjointed domains. Even when we view the supervening entities as somehow composed of those from the lower level, supervenience could still be holistic in nature; there might be a many–many relation between the supervening and base predicates or facts. For example, the totality of my beliefs might be supervenient upon and token identical with some complex of neurological states, even though there was no one neurological state that realized only my belief that *P*. Supervenience does not guarantee that individual predicates match up in a one-to-one fashion as the standard notion of reduction requires.

So there are overwhelming reasons to think supervenience does not *entail* reducibility. A host of conditions besides supervenience must be met, and it is an empirical issue whether they are in each case. A nonreductive unity of science thus remains possible.

Explanatory Impotence?

I turn now to a second batch of doubts about the nonreductive unity of science I advocate. The basic intuition goes this way: If higher-level entities and their characteristics are composed from and dependent upon lower-level entities and their traits, then they cannot be real causal factors—they are at best epiphenomenal. Jaegwon Kim has forcefully pressed this idea, and below I look at two such arguments.

Nonreductive unity holds that there can be successful higher-level explanations, and yet those explanations may not be captured in lower-level terms because of, among other things, multiple realizations. Kim doubts that this claim is coherent. If higher-level kinds provide genuine explanations and are multiply realized, then he claims a dilemma arises. Either the various lower-level states realizing that higher level-kind themselves constitute a kind or they do not. If they do not, then obviously the higher-level kind is nothing of the sort. If they do form a kind, then we have reduction after all and multiple realizations are no problem. So either the special sciences describe no kinds—in which case they are explanatorily suspect—or their realization does make a kind and thus reduction ensues.

To evaluate this argument we need an account of natural kinds. No such account has universal assent, but I take the following to be some widely shared claims about natural kinds: Natural kinds are those categories that can play a central role in explanation. They are the categories that laws relate. They are the categories that are projectible in Goodman's sense—they allow us to successfully generalize about new cases. If these claims are reasonable, then surely many higher-level kinds are natural kinds. "Fitness" and "signal sequence" are obvious examples, for we can explain with them, they are projectible, and they do ground generalizations (albeit qualified ones). Yet they are clearly multiply realized in physical terms. So multiple realization per se cannot show they are not natural kinds.

What about Kim's claim that if they are kinds, then their total set of realizations must be as well? In one sense we can grant this claim, but it will do nothing for the reductionist project. In another sense this claim is quite implausible. The claim is implausible if we are talking about actual realizations, for reasons we saw above: actual realizations do not suffice to capture the modal claims that natural kinds ground. If we move to possible realizations instead, then we may grant that this set is a kind if we take the "possibility" broadly enough. But now the second problem sets in, namely, supervenience does not ensure that we can even in principle describe all the possible realizations. So reducibility in the interesting sense does not follow. Higher-level categories like fitness or signal sequence give us a way of talking about those possible realizations without naming them. That is what natural kind terms do. The disjunct of their physical realizations serves no such function—unless we construe the possibility so broadly as to make it beyond the grasp of finite agents like ourselves.

Kim has another argument in the same spirit. Nonreductive unity leaves the higher-level sciences causally otiose or it postulates nonphysical forces. Kim argues the point in terms of the mental and physical, but his arguments are perfectly general. If some mental state *M* causes later states of my body, then what is the causal role of the physical state *P* that realizes *M*? Is it not itself sufficient to cause my behavior? If it is, then the mental properties of my brain state play no role; the mental is not causally active qua mental. If the realizing state *P* is not sufficient, then we are stuck with nonphysical forces.

Kim's insistence on interpreting every claim as one about properties leads us astray again, as it did above in the question-begging formulation of supervenience. "Properties" as Kim uses them are types, not tokens. So he is able to talk of mental properties and assume the antireductionist will take them to be different from physical properties because of multiple realizations. Then their role seems mysterious. Yet we can avoid the mystery by eschewing the commitment to properties in the philosopher's sense altogether. *Token* mental states—or any token of a higher-level state—are realized by or identical to token states at the next lower level. This is not incompatible with the claim that mental kinds may not be identical to some physical kind. However, token identity is all we need for causal efficacy. Being a signal sequence in a particular case is having a certain biochemical structure, and that structure is causally efficacious. "Being a signal sequence," taken as referring to the trait instance, is a causal factor, and it is so without contravening the physical. The mystery is no longer.

Two Opposing Views

Supervenience does not entail reducibility nor does it make higher-level properties causally inert. Yet it seems to many that supervenience must somehow allow us to explain everything in lower-level terms, even if reduction is not an option. For example, individualists who do not advocate reduction claim nonetheless that every social event "can be fully explained individualistically" (Danto 1973). Similar moves are made by the reductively minded in biology and psychology. Rosenberg, for example, acknowledges that biological kinds and physical kinds do not match up in the way required in standard accounts of reduction but nonetheless asserts that "all biological phenomena could be best and most fully explained by theories about their chemical constituents" (Rosenberg 1985, 63). Similarly, critics of Gould's punctuated equilibrium approach to evolutionary theory admit that facts about species selection cannot be reduced to facts about organismic selection but claim the latter can fully explain species-level phenomena. Stich (1985) tries a similar tack in arguing for his variant of cognitive psychology: human behavior probably can be fully and best explained by a purely syntactical account, even though the prospects for reducing folk psychology to a syntactical theory are dim.

None of these authors provide any detailed arguments establishing that explanation can be split from reduction as their claims suppose. However, the general idea seems to be this: the higher-level facts and processes to be explained are obviously constituted from lower-level facts and processes and do not occur independently of them. Lower-level explanations thus cannot help but explain higher-level events and do so more fundamentally. Higher-level theories may be partially explanatory and irreducible, but as the science in question further develops it will be clear that lower-level accounts fully and best explain all that needs explaining.

Ironically, many philosophers have argued for the opposite conclusion, namely, that we can explain nothing about higher-level events in terms of the phenomena that realize them and on which they supervene. Rejecting such supervenience explanations for psychology, Dennett (1969) claims that "one could associate each true sentence of the mental language with a sentence that catalogued as exhaustively as possible the entire physical state of the person or persons in question, but this would explain nothing." Similar conclusions are reached by Garfinkel (1981), Putnam (1981), and Wimsatt (1976).

The arguments for the claim that supervenience explanations are in fact nonexplanatory take several different routes. One common line argues that supervenience explanations provide irrelevant detail and thus do not explain (Putnam 1981). Garfinkel (1981) argues that supervenience explanations do not describe events necessary for the occurrence of the higher-level event in question and thus do not explain. Supervenience explanations are also criticized because they provide so much detail that the resulting explanations are impossible to grasp. Finally, Kitcher (1984) has argued that lower-level theories in biology, for example molecular biological accounts, are inadequate because they make what is homogeneous "look heterogeneous."

Of course, both claims threaten the nonreductive picture I have been advocating. If supervenience explanations are fully adequate, then the special sciences lose much of their autonomy; if supervenience explanations explain nothing, then the search for integrative, interlevel theories is misguided. In the remainder of this section I set up the issues with more care, and then turn in the final two sections to reject both attacks.

Explanation, at least at its best, is done by (well-confirmed) theories and so what is at issue is whether lower-level theories can fully explain higher-level phenomena when: (1) a confirmed high-level theory is not reducible to the lower-level theory, (2) the higher-level phenomena supervene on lower-level phenomena, and (3) higher-level phenomena are realized in their lower-level counterparts. More formally, let *Tw* and *Tp* be fully developed theories such that *Tw* refers to entities of kind *W* and *Tp* to entities of kind *P* and every entity of kind *W* is composed of entities of kind *P*. Now the thesis under dispute is

> Every *W* and its properties can be fully explained by *Tp*—(on the assumption that the following three provisos hold:
> (1) Each essential predicate of *Tw* is not connected in a lawlike manner with some simple or conjunctive predicate in *Tp*, that is, *Tw* is not reducible to *Tp*;
> (2) once all the facts expressible in *Tp* are set, then so too are all the facts expressed in *Tw*, i.e., the latter facts supervene on the former; and
> (3) every *W* is either (a) identical to some *P* or sum of *Ps* or (b) is constituted solely from some *P* or *Ps*, that is, the higher-level entities are realized by their lower-level counterparts.

The two versions of (3) represent two different ways of understanding the idea that something is realized in something else. One natural rendering of this idea is that the two things are token identical—to say that some segment of DNA realizes a particular gene means that in this instance this particular gene is that sequence of bases. However, the notion of token identity may not always be well defined. It is often unclear, for example, with what physical token we should identify complex objects such as universities. The second construal of "realized in" is meant to allow for such situations, leaving the defender of supervenience explanation to define the supervenience base as broadly as necessary to overcome fuzzy cases. Thus (b) represents the less-controversial rendering of "realized" as "solely constituted by."

Before considering whether supervenience and realization ensure that *Tp* is fully explanatory, one further clarification is needed: what constitutes an adequate explanation? Obviously this topic is enormously complicated and cannot be discussed in detail here. I will rather outline an account that draws on a large body of current research and is yet neutral on many controversial issues.

Following the work of van Fraassen (1980), Garfinkel (1981), and Achinstein (1983), we can consider an explanation as an answer to a why-question. Questions are not answered simpliciter but must be specified relative to certain parameters determined by context. Among those parameters are at least the following.

1. *The Contrast Class.* A simple why-question is potentially ambiguous, for the answer can vary depending on how the possible answers are sorted. Take, for example, "why did Gridley expire?" Answering the question involves saying why the topic of the question—Gridley's expiration—arose rather than some other possibility. But there are numerous ways to conceive those alternatives. How we contrast the topic, that is, in what contrast class we place it, will affect our answer. "Why did Gridley expire?" may get different answers depending on whether the contrast class is {Gridley expires, Gridley recovers} rather than {Gridley expires, Jerome expires}.
2. *The Relevant Kind of Answer.* After the contrast class is set, answers may still vary depending on what is taken to be relevant. When we ask for the cause of Gridley's death, different answers are appropriate depending on the context. For example, heart failure, circulatory stress, and a genetic defect could all be listed as the cause of death due to sickle-cell anemia.
3. *Background Theory.* Even after the contrast class and relevant kind of answer are set, questions still need to be further specified. Questions with different background assumptions require different answers. Even if we are looking for the immediate cause of death, for example, the relevant answer will vary with different theories of the causal process.

An explanation is thus an answer to a question that must be specified according to contextual parameters. This account breaks with the Hempelian tradition that looked for a purely formal explanatory relation between the explanans and explanadum statements. That quest has run up against numerous counterexamples. The contextualist model shows why that should be: explanation is not a formal relation between the explanans and explanadum but rather depends crucially on contingent, empirical facts about the context. Most, if not all, of the standard counterexamples result from changes in contextual parameters.

Using this outline, the explanatory power of a theory can be evaluated along two lines: (1) by its ability to answer any given, fully specified question and (2) by the number of relevant questions it can answer. Much of the standard literature on explanation concerns the first dimension and seeks to determine when an answer to a question fully accounts for the facts it cites. Philosophers disagree on the criteria for successful explanations, though any account will have to make the citing of causes as one paradigm sort of answer. However, the arguments that follow presuppose no very specific view about these issues.

Evaluating questions by the number of questions they answer (however they do so) will play an important role in my argument. Once again, there is no worked out account for this aspect of evaluation either, due this time to a lack of discussion, not conflicting accounts. The work of Belnap and Steele (1976) on the logic of questions and answers provides some material for individuating questions. Nonetheless, I know of no acceptable precise quantitative way of measuring this second aspect of evaluation. However, for the arguments that follow I only need assume that we can make rough judgments that one theory answers some questions another cannot. Further precision will be unnecessary.

Thus, the basic questions I will be pursuing can be more perspicuously put this way: do supervenience and realization ensure, even when reduction fails, that a fully developed lower-level theory *Tp* can adequately answer the relevant questions about each higher-level entity? Or is it the case that *Tp* can tell us nothing, something we saw earlier that Putnam and others claim?

Explanation without Reduction

Supervenience explanations are inadequate, I want to argue, because they are incomplete and partial. A "partial explanation" may sound like a contradiction in terms, but it is not, at least not on the account of explanation sketched earlier. Since an explanation is an answer to a question, a theory may be able to fully answer some questions without thereby being able to answer other relevant questions. In such a case, the theory only partially explains. Lower-level theories, I want to argue, may be partial explanations in just this sense. Questions must be specified along at least two parameters: the contrast class and restrictions on relevant answers. Contrast classes are often composed of natural kinds applying to the event to be explained. While we could specify the relevant contrast class by referring only to particular individuals (Why did John die—in contrast to Jerome, Adolph, etc.), we can and often do ask why did this kind of event occurs in contrast to other kinds of events. Similarly, a natural restriction on possible answers limits the latter to those that connect the event in question to other kinds of events—what kind of events usually cause this kind of event, what kind of effects does this kind of event have, what kind of events might have brought this event about, and what kind of functional role do events of this kind have. In short, many questions require us to relate the event in question as a kind to other kinds of events, structures, and the like.

Supervenience explanations, however, cannot answer many questions of this sort for a simple reason: they do not have the relevant kind terms referring to the event in question nor to the other higher-level events to which it may be related. As defined here, supervenience explanations proceed by invoking lower-level kinds and laws applying to the component parts of the higher-level phenomena to be explained, not to the higher-level phenomena themselves. Furthermore, since we are assuming reducibility fails, there will be no way to equate the higher-level kinds with the conjunction of lower-level predicates describing the supervenience base involved (if there were such a conjunctive lower-level predicate that was coextensive with some higher-level kind, then the conditions for reducibility would be satisfied). So a lower-level theory will have no simple predicates referring to kinds of higher-level events and no substitute complex terms either. We will thus lose access to the network of connections between the particular kind of macroevent to be explained and the relevant kinds of causal antecedents and effects at the macrolevel. As a result, lower-level theories will necessarily leave a great many questions unanswered.

Failing to answer such questions is no trivial inadequacy. Much current social explanation proceeds by tracing out connections at the macrolevel. Macroeconomic theories, for example, relate sectors, institutions, and other macrolevel entities and, as we will see in the next two chapters, there seems to be little prospect of cashing out these explanations in purely microeconomic terms (see also Nelson 1984). While historians often explain by tracing out the actions of important individuals, they also explain by relating large-scale events, like revolutions, to structural and institutional precedents and their typical effects. Mendelian genetics represents a similar situation; it relates kinds that have no obvious equivalent in the biochemical structure that realize them. Because fitness is apparently supervenient on but not reducible to physical properties, evolutionary theory likewise will explain largely in terms of higher-level kinds (see Sober 1984). As suggested earlier, cell biology may be in a similar situation in relation to molecular explanations. Because supervenience explanations would deny us all these higher-level kinds, they obviously must leave many questions unanswered. And for that reason they will be at best partial explanations.

We are now in a position to make sense of and more rigorously defend the critic's complaints that supervenience explanations (1) do not bring out the salient features, (2) make homogeneous events at the macrolevel look heterogeneous, and (3) fail to explain because they do not cite factors necessary for the event to be explained. Each objection points to the fact that lower-level accounts cannot make use of important kind terms to explain higher-level events. Lacking the relevant kind descriptions, lower-level explanations will miss important causal connections to other kinds of events—in short, they will miss causally relevant properties at the macrolevel. Because they cannot capture macrolevel kinds, supervenience explanations will provide disparate explanations of events that do have a common macrolevel account. Similarly, because lower-level theories lack the appropriate kinds, they cannot make important counterfactual judgments of the form "This kind of macroevent would have occurred even if its actual constituent components were not present." It is these counterfactual claims that Garfinkel seems to have in mind when he rejects supervenience explanation because it does not provide necessary conditions. While I will argue in the next section that these problems do not show supervenience explanation to be inherently inadequate, critics are nonetheless pointing to real problems.

The arguments of this section rest on two crucial assumptions that should be made explicit and defended: (1) that the questions that higher-level theories can answer are important—that they are questions that an adequate account ought to answer and (2) that higher-level theories can indeed adequately answer some questions. Those assumptions might well be challenged. Let me begin their defense by starting with the former.

Any scientific theory, no matter how good, will probably leave some questions unanswered. Aside from those that are simply unanswerable (because they have false presuppositions, for example), there will always be unanswered questions so long as we have vocabulary that goes beyond that of the theory in question. So why think that supervenience explanations are inadequate because they cannot answer questions posed by higher-level theories? Why must they answer such questions?

Citing certain questions as ones that ought to be answered, or as important, involves in part a value judgment. However, such judgments are probably endemic to science and scientific progress; Kuhnian revolutions and thus normal science involves decisions about what ought to be explained. There are, however, good reasons to think we will find the questions of many higher-level theories worth answering, for they involve many enduring, if not intrinsic, human interests. For example,

1. Our very concept of a person—which is crucial to our moral concerns—seems to presuppose that we use psychological concepts associated with rationality, choice, and the like.
2. The ideological struggles that have motivated much of modern history are about institutions such as forms of government and kinds of social organization.
3. Our interest in understanding ourselves must be partially in terms of how our species has evolved due to the forces of evolution.

But explanations in terms of psychology, institutions, and biological entities are all higher-level in nature, and the list could easily be expanded. We have natural reasons to think that higher-level questions are worth answering.

Furthermore, lower-level accounts frequently presuppose the importance of higher-level questions. We saw in chapters 2 and 3 how this can happen in the social sciences, and chapters 6 and 7 will provide further evidence. Similarly, we saw in the last chapter that many molecular biological explanations implicitly invoke nonmolecular truths because they are really functional in nature. Molecular biology is rife with apparently essential functional terms such as "receptor," "signal," or "messenger." However, such functional descriptions appeal to roles within larger systems, in short, they appeal to higher-level explanations, in this case explanations involving cells and organelles. In addition, much progress in providing lower-level explanations typically comes from answering higher-level questions. Higher-level explanations provide kinds of groupings that make visible mechanisms on the lower levels that would otherwise be lost in a mass of details. Heuristically, answering higher-level questions may go hand in hand with developing lower-level theories. In short, there are good reasons to believe that lower-level questions often are not separable from those at higher levels.

The second implicit assumption I need to defend is that higher-level theories ever really explain. A more radical question eliminator might not just deny that higher-level theories are important but also take the more radical stance that they are in principle unanswerable. Those who argue that folk psychology is a radically false theory might be read as denying that higher-level psychological theories explain at all; some methodological individualists argue for a similar conclusion about sociological explanation. If higher-level theories do not really explain, then they ought to be eliminated and pose no threat to supervenience explanation.

Does such eliminativism undercut the arguments of this section? If it is supported by reasons that hold only for a particular domain, then eliminativism constitutes no problem for the general position defended here. And, of course, eliminativism in philosophy of mind is typically supported on grounds specific to psychology, for example, problems associated with folk psychology. However, defenders of eliminativism sometimes, either explicitly or implicitly, give arguments that would show that any higher-level theory is nonexplanatory. Such arguments are much more serious threats to the position I have been defending.

Methodological individualists frequently imply that any macrolevel theory is inadequate because lower-level theories deal with what is more fundamental and thus higher-level theories are always incomplete and inadequate. In a similar vein, Elster (1985) has argued that lower-level accounts are to be favored because (1) they reduce the time lag between cause and effect and thus reduce the prospect of spurious explanation and (2) are more detailed and thus to be preferred.

Two of these individualist arguments simply beg the question. The claim that lower-level explanations are to be preferred because they deal with more fundamental objects assumes that lower-level explanations are themselves adequate. Similarly, appealing to the superior detail of lower-level explanations assumes that this detail suffices to explain. But these assumptions are just the claims at issue. Higher-level theories can potentially answer a wide range of questions that supervenience explanations cannot. The eliminativist must thus do more than appeal to the existence and virtue of lower-level accounts: she must show that any theory employing higher-level kinds is false.

Elster's other claim—that lower-level explanations are to be preferred because they avoid spurious causation—is potentially more promising but nonetheless mistaken. Heuristically, one mode of inquiry may in fact tend to produce more errors than another, but that says nothing about whether the respective theories explain when they are adequately developed. A theory that recommends "look for macroconnections" might lead to errors more often that some alternative heuristic, but that does not entail that there are not true macroconnections and explanations. Furthermore, lower-level theories do not, so far as I can see, necessarily have anything to do with shortening the gap between cause and effect. Presumably when some whole causes some event, the action of its collected parts is simultaneous with the action of the whole. To assume otherwise is to envision a very bizarre metaphysics indeed. But if the collected parts act simultaneous with the action of the whole, then neither level of explanation better reduces the gap between cause and effect.

A second route for supporting a general eliminativism can be found in Churchland's (1978) arguments against folk psychology. Folk psychology, he argues, is a radically false theory because, among other things, it cannot explain mental illness, the nature of creative imagination, and certain perceptual abilities. At work here is the idea that an incomplete explanation is no explanation at all. If that claim were true, then pure higher-order theories (higher-level theories that make no reference to lower-level phenomena) would be inherently inadequate, for there is much they cannot explain: any question calling for microstructure and the mechanisms that realize higher-level events, for example.

However, eliminativism motivated in this way runs into two problems. For one, it is wrong to assume that a theory that only partially explains is inherently inadequate. A theory may be incomplete in that it answers only some relevant questions; nonetheless, its failure to answer all relevant questions does not mean that it cannot adequately answer any question, and thus does not show that a pure macrolevel theory is nonexplanatory. Insofar as eliminativism relies on an all-or-nothing picture of explanation, it is misguided.

While earlier I granted that higher-level theories could not answer certain questions and were to that extent incomplete, not every higher-level theory has such a problem. If we distinguish pure from mixed higher-level theories (the latter allowing some reference to the component entities described by lower-level theories), then only theories of the pure variety are excluded from answering questions about the mechanisms that realize macroevents. So even if an incomplete explanation were no explanation at all, this argument would not motivate an eliminativism about mixed higher-level theories. But the defenders of supervenience explanations need an eliminativism of the latter sort, for they claim that no theory employing higher-level kinds is adequate.

Another common attack on the higher-level sciences turns on the qualified nature of their putative laws. Laws in the social and biological sciences are often qualified with ceteris paribus clauses. Is this perhaps true of all higher-level explanations and thus a reason to deny that those explanations are genuine? I do believe that all or at least most laws in the social science hold only ceteris paribus. However, this fact carries no weight by itself for all or most laws in physics are likewise qualified by ceteris paribus. Since I have pursued this argument in detail elsewhere (1996), I will not consider it further here.

So the main claim of this section remains unscathed; supervenience explanations are seriously inadequate because they cannot handle numerous questions concerning kinds and counterfactuals. Those problems cannot be skirted by any *general* argument for eliminativism, for those arguments are quite weak. Supervenience does not make higher-level explanations redundant. And, as a result, eliminativists who motivate their own position by appeal to supervenience explanation do so illegitimately.

Do Lower-Level Theories Explain at All?

In this last section I want to argue against those philosophers—Putnam, Garfinkel, Wimsatt, or Dennett—who hold that lower-level counts do not explain. Below I first use the partial account of explanation developed in this chapter and earlier ones to argue that supervenience and realization can ensure, at least under some conditions, that lower-level theories do explain higher-level events. I then criticize the arguments claiming that supervenience explanations are no explanations at all.

Given what we said earlier about explanation, it should be fairly obvious why lower-level theories explain higher-level events. A direct route to this conclusion comes if we grant that explanations are not just answers to why-questions but may also answer how-questions as well. Then lower-level theories will clearly answer some questions and thus be at least partially explanatory if supervenience and realization hold. For example, if my question is something like "How do human liver cells recognize and internally transport macromolecules?" then an account of the molecular events upon which transport and recognition supervene will answer my question. Lower-level theories cite realizing facts and thus can answer my question. Denying them any explanatory value consequently goes too far.

Note that this conclusion holds even if reduction fails. As we saw in chapter 4, the chances are good that many general cellular processes can be realized by indefinitely many molecular mechanisms and that cellular biology is as a result irreducible to a purely molecular account. Statements about cellular recognition and intracellular transport may be irreducible for such reasons. Nonetheless, if our question concerns how some particular cells—rather than cells in general—recognize and transport, then we have a question that can be given a determinate answer. Supervenience holds even if the supervenience base is not coextensive with the supervening facts as kinds or types. So we are guaranteed that the lower-level theory can explain how these particular cells act even if it cannot provide one general reductive explanation of how all do so.

My argument so far has turned on counting correct answers to how-questions as genuine explanations. And that seems a reasonable assumption, given that a great deal of scientific inquiry can be plausibly construed as seeking to answer such how-questions (how do cells do x, y, and z, how did the universe originate, etc.). Nonetheless, we can restrict scientific explanation to why-questions without loss, for the content of how-questions is framable in suitably construed why-questions. Questions, remember, must be specified for contrast class and a relevant range of answers. So if we ask why pancreas cells secrete insulin rather than a . . . z, a variety of answers are possible: answers citing facts of embryology, the appropriate evolutionary history, the function of the pancreas in the overall biology of the body, and the molecular mechanisms realizing the function of the pancreas. But the latter answers tell us in effect how the cell operates.

Even if the citing of microstructure alone is explanatory, there are still stronger reasons to believe that lower-level theories do explain higher-level phenomena. Although explanations by appeal to microstructure might be controversial, surely explanation by appeal to causes are not. However, if we make the further reasonable assumption that singular causal claims are extensional, then lower-level theories will be able to answer some why-questions about higher-level events. To see why this is so, imagine that our lower-level theory *Tp* completely describes the total state of the component parts—call it S_1—realizing some higher-level entity *W* in a given state *Ws*. If *Tp* is in fact an adequate theory, it will then also be able to explain the cause (which would of course be complex) of S_1 coming to exist and can thus answer the question "Why *S*?" with a causal answer. However, if S_1 is token identical with *Ws*, then to answer "Why S_1?" is equally to answer "Why *Ws*?". Thus on some minimal assumptions, supervenience and realization ensure that *Tp* can give causal explanations of higher-level events.

Can we make a similar argument for explanatory adequacy of *Tp* if we interpret "realization" in the weaker sense of "constituted by"? The crux of the issue is this: if we identify the causes of all the components making up some particular event, have we thereby identified the cause of that event itself? Intuitively, the answer seems affirmative. Once we have explained what caused the subsystems of the space shuttle to function as they did in the context of flight L51, we have explained why the shuttle exploded. While the event in question may have no single causal explanation as a kind (explosions can happen in definitely many ways), the cause of this particular explosion seems fully specified once we fully specify the causes of the subsystems bringing it about. In short, it is prima facie plausible that constitutive identity preserves substitution in singular causal claims. If so, supervenience and realization, even on its weaker interpretation, guarantee that *Tp* can provide singular causal explanations of higher-level events.

What has motivated philosophers to deny that lower-level accounts explain macrolevel phenomena? Discussing reductionism in psychology, Putnam (1981) asked whether we could give a solely microstructural or lower-level explanation of something so simple as why a peg fits through a hole in a board. He argues that even if we could completely describe the position and velocity of every particle and thereby deduce from that information that the peg would not fit through the board, we still would not have explained the latter fact. The reasons he gives are common to those who criticize supervenience explanation: the lower-level account provides irrelevant information and hides the salient information under a mass of detail.

I think Putnam is simply wrong in claiming that lower-level explanations necessarily give irrelevant detail. Consider the example he provides. For the specific peg and board in question, their rigidity and shape just is having molecules in a certain arrangement. The peg fails to fit when certain molecules in a particular arrangement and relative velocity come into contact with others in a certain arrangement and velocity. Following our argument above, if a description of the rigidity and shape answer the relevant question, then so does citing the relevant microstructure, since the former macroproperties in this case are realized by the microstructure in question.

When Putnam says that the microstructure is irrelevant and thus nonexplanatory, he appeals to the fact that macroproperties, rigidity and shape in his example, can be realized by indefinitely many different microconfigurations. The microstructure is then irrelevant in the sense that it is not necessary for macroproperties used in the explanation. Garfinkel, who argues along lines similar to Putnam, puts the point this way: "the occurrence of the specific microstate *X* was not necessary for the occurrence of the qualitative outcome, and hence it is counterexplanatory to include it in the explanation" (1981, 63).

Putnam and Garfinkel are correct that the microstate is not necessary, at least if higher-level predicates are multiply realized. Supervenience only guarantees that the microstate is sufficient for the macrostate. However, does this show the explanations invoking only the microstate to be nonexplanatory?

I think not. The facts about the microstate need not be necessary to the higher-order phenomena in order for those facts to answer some relevant questions. If I ask "why am I dying?" and the doctor cites the particular cellular facts of my disease, he has given me a real enough explanation: that I could be dying some other way hardly makes his answer nonexplanatory. And, of course, there is a necessary relation between the macrostate and its supervenience base: given the microstate in question, the relevant macrostate results necessarily.[1] While many questions are not answered by appeal to lower-level facts, that does not hold for every kind of question. We sometimes ask for explanations of *tokens*—specific events—by appeal to other tokens. But a particular peg's rigidity, unlike rigidity in general, is identical with or realized in a particular microconfiguration. Thus, when our questions are about tokens rather than kinds, supervenience explanations can provide the answer. So lower-level accounts can answer some questions; Putnam and Garfinkel are thus wrong to call them nonexplanatory.

There is surely something to Putnam's claim that supervenience explanations do not bring out the salient features. Macrolevel properties are often more accessible and manageable than their lower-level realizations—even though the two are in fact identical. I am not sure, however, what this fact shows about lower-level explanations. First of all, there is no reason to think that every lower-level explanation must hide the salient features. Not all microlevel accounts are as complicated as Putnam's example, where we must make a very large number of calculations. Frequently the lower-level details are not so complex that we cannot see how they explain—in fact, we have innumerable such explanations in undergraduate textbooks, for example, molecular accounts of specific genes, membrane function, protein transport; or explanations of the digital electronics realizing computer programs. And in some cases, it is only when we see the lower-level realization that the macrolevel properties become salient. Thus the pragmatic difficulties of some lower-level explanation are not inherent in the genre.

Furthermore, even if every lower-level account were always so complex that we had difficulty in seeing the overall picture, that would not show that they are not explanations—any more than the fact that some mathematical proofs are too long and complicated for one individual to follow shows that they are not proofs. While there is no doubt a subjective sense of "explains" that equates explanation with understanding, there is also an objective sense of "explains" that I would argue is both primary and irreducible to a subjective notion. The account of explanation sketched earlier built in several contextual factors, including the background information that often determines what a particular audience can grasp. But even if that information is the information of a specific group, there remains room for an objective sense of "explains." Once the contrast class, relevant kind of answer, and given background information are specified, then it is a factual, objective question whether or not some proposed statement answers the question at hand. In short, we can grant (1) that there is a subjective sense of "explains" and (2) that pragmatic factors are involved in explanation and yet still deny that pragmatic difficulties show lower-level explanations are nonexplanatory.

My defense of supervenience explanation is, of course, still a qualified one. Supervenience may often hold only globally: although all the facts expressed by a lower-level theory may fix all the higher-level facts, it may still be the case that when we restrict the lower-level facts to the more narrow set of facts realizing a given macrophenomena, then supervenience fails. However, if supervenience fails, then a lower-level explanation of the entities composing some whole will not thereby guarantee that we are also explaining the whole in question. For example, two computers in the same machine-state, it might be argued, need not have the same program, for specifying the program depends on how inputs and outputs are interpreted. Program states are then not supervenient simply upon machine states, and as a result a complete account of the latter does not explain the former.

A second condition that must be met concerns how the lower-level facts are specified. If the lower-level theory *Tp* describes its domain in functional terms, then there is the possibility that its descriptions presuppose higher-level facts. As we saw in the last chapter, an individualist social theory that explained some large-scale social event by use of predicates such as "teacher," "union leader," "prisoner"—in short, that referred to roles within institutions—would arguably be presupposing higher-level facts. Since these descriptions refer to roles within an institution, they implicitly presuppose a host of facts about those institutions. Hence what would look like a purely lower-level explanation would in fact not be. There is, of course, no reason to think that every lower-level explanation must commit such sins. Nonetheless, lower-level theories cannot claim to explain unless their descriptions do not implicitly presuppose higher-level facts.

If such conditions are met, then supervenience explanations are not only explanatory, they are in at least one sense fundamental. Assuming that higher-level theories supervene upon their lower-level counterparts, then the same holds for explanations: once the lower-level explanation is set, then so too are all adequate higher-level explanations. In other words, supervenience explanations determine or fix explanations at other levels. Lower-level theories, regardless of reducibility, not only explain but also have a kind of primacy.

6

Individualism, Explanatory Power, and Neoclassical Economics

Individualist approaches are often defended on the grounds that they provide the *best explanation* of social phenomena. In what follows I look at these defenses, examining what appeals to explanatory power can legitimately do. My specific focus throughout shall be on one specific version of the individualist program, namely, neoclassical economics. Despite its problems, the neoclassical program seems to many to be the best thing going in economics because of its ability to explain a wide variety of phenomena, both economic and social. That explanatory power, many economists argue, more than compensates for other inadequacies of the neoclassical tradition. I argue in this paper that they are wrong.

Though my focus is on neoclassical economics, much larger issues are of course lurking in the background. The main competitors to the neoclassical program, especially when it comes to its individualism, are institutionalist, Marxist, and other approaches that want to build in social structure from the start in economic analysis. And insofar as these approaches defend themselves by attacking methodological individualism, my topic also shades into the general individualism–holism dispute. The arguments of this chapter will thus tie into the larger issues discussed in previous chapters. However, my primary concern is with the value of explanatory power as a defense of individualism. Since explanatory power attaches most concretely to specific theories, progress will come most directly by examining specific individualist theories in detail. Hence the concentration on neoclassical theory.

This chapter is organized as follows. After further clarifying the preliminary issues, I discuss the place of explanatory power in confirmation. Appeals to explanatory power—generally known as inferences to the best explanation (IBE)—are often taken to be the crux of scientific inference; in other words, IBE is thought to be a foundational inference rule. I argue that it is no such thing. Appeals to explanatory power rest on substantive, contingent, and often implicit assumptions to do their work. Moreover, theories with great explanatory power can nonetheless be poorly confirmed. So quick appeals to the explanatory power of neoclassical economics or individualism more generally will not do. Rather, we need to assess the substantive notions of explanation involved and to weigh explanatory power against other empirical virtues. The third section begins that process. It looks at the idea that explanation comes from unification, particularly unification as spelled out by Kitcher (1989) in terms of "argument patterns." This picture of explanatory power quite nicely describes the virtues economists see in neoclassical theory. However, I argue both that unification is a suspect notion of explanation and that, commensurate with my general claims about IBE, this notion of explanatory power rests on important substantive assumptions about the economic and social world—assumptions that are empirically suspect. The following section turns to explanatory power taken as the ability to cite causes. Again I argue that neoclassical economics has superior explanatory power only given some quite dubious empirical assumptions. Since accounts of explanation usually fall either into the unificationist or causal camps, the upshot is that the neoclassical program and its individualism gain little credibility by appealing to explanatory power.

Some Preliminary Issues

Before turning to my main arguments, we need to clear some ground. In particular, we need to be clear about our target—"neoclassical theory"—as well as some lurking presuppositions of the claim that neoclassical theory best explains and thus that individualism wins the day. I start first with neoclassical theory.

I doubt that there is any *one* thing called neoclassical theory. Attempts to list its essential postulates always run aground of the fact that, like most disciplines, theories in economics get different interpretations and emphasis depending on the task at hand. So to describe my target, I can at best point to some core principles grounding the family resemblance between different applications. In particular, neoclassical theory is usually committed to some part of the following:

1. Economic outcomes must be explained as entirely the result of individual choices.
2. Those choices are rational.
3. Rational choices are those that maximize self-interest given constraints.
4. Choices are coordinated by markets.
5. Markets are best understood by focusing on full competition and equilibrium outcomes.
6. Full competition entails complete prices, full information about prices and technology, price-taking behavior by firms and consumers, and free flow of resources to new uses.
7. Markets produce efficient outcomes—firms equate marginal revenues to marginal products and so on.
8. Incomes are returns to factors.

When economists extol the explanatory virtues of neoclassical theory, they are claiming that these general ideas provide the best explanatory strategy—that fleshing out these postulates for the case at hand is the most successful route to explanation and that this is powerful evidence in its favor. It is this broad claim that it my target.

Two crucial presuppositions are hiding in the claim that neoclassical theory best explains and thus supports individualism: (1) that neoclassical theory is individualist and (2) that individualist theories of the neoclassical variety compete with—in short, are incompatible with—their nonindividualist counterparts. Both assumptions are widely made and yet both are open to question.

In many of its formulations, neoclassical theory is not a thoroughly individualist theory, where an individualist theory is one that makes no essential reference to collective entities. Neoclassical theory most obviously contradicts its individualist ideology in the traditional theory of the firm. Firms are typically treated as black boxes and given no more content than comes from the standard production function and profit maximization. How individual behavior brings about this production function and profit maximizing is left unsaid. In short, "firm" is taken as a primitive predicate, one not further analyzable in individualist terms. So standard formulations of neoclassical theory make essential reference to social entities.

Moreover, firms are not the only way neoclassical theory violates its commitment to individualism. Aggregate concepts show up elsewhere. Consumer theory invokes aggregate concepts when it explains in terms of households or of aggregate market demands. And consumer theory, as do other parts of neoclassical economics, invokes the notion of representative agents. These "representative agents" are really aggregate social entities that are treated as if they were individuals with preferences, knowledge, and the like. These aggregate attributions, however, generally cannot be derived from the standard neoclassical postulates about individual human agents.

Of course, I see nothing wrong in such explicit appeals to social or collective entities. However, since I want to use neoclassical theory as a paradigm individualist theory, it is important to note these exceptions. They are, I think, recognized by at least some economists as clearly inconsistent with the traditional commitment to individualism. For example, much new work on the theory of the firm aims to explain how firm behavior results from the actions of maximizing individuals; the traditional account is explicitly rejected because it violates the maxims of individualism. So the commitment to individualism runs deep. Nonetheless, we should keep in mind that even if neoclassical economics were entirely successful, that theory often explicitly invokes social entities in its explanations and is to that extent ambiguous evidence for individualism.

Yet another important presupposition is at work in the claim that neoclassical theory best explains: namely, that neoclassical explanations are incompatible with those of its competitors. If the explanations were complementary—as are biological and biochemical accounts of disease, then claims that one is the better explanation would border on incoherence. This problem is exacerbated by the fact that individualist, neoclassical accounts often use different vocabularies and explain at different levels than do their competitors, increasing the odds that they simply describe different parts of reality or divide the same reality in different ways.

The ideal way to determine when and where individualist theories actually compete with social accounts would be to reduce the latter to the former. Then we could directly compare their predictions and explanations. However, I am skeptical that any such reductions will be forthcoming and am even more certain we do not have them now. That leaves us asking case by case whether some social explanation of some specific event when cashed out in individualist terms entails individual behavior incompatible with the preferred individualist theory, which in this case is neoclassical economics. For example, we would need to look at specific institutionalist social explanations of the firm and determine whether they entailed individual behavior incompatible with standard neoclassical accounts. Only then would we have a case of competing theories and only then would it make sense to claim the neoclassical account better.

No doubt neoclassical theories often do compete with their rivals. Yet I will argue below that they just as often implicitly presuppose or take as given the phenomena that social theories make central. That means neoclassical theory in those cases does not compete in the way necessary for claims about the best explanation to make sense. To that extent, the appeal to explanatory power will again fail to support neoclassicism and the individualist program.

Explanatory Virtues and Confirmation

To evaluate neoclassical economics by its explanatory power requires that we first get clear on the general role of explanatory power in confirmation. This section attempts to do that. I argue that IBE depends upon substantive, often domain-specific assumptions for its force. I also argue that on this understanding of IBE, the most explanatory theory is not always the best confirmed theory. These general morals will play a crucial role as we look at the explanatory virtues of neoclassical theory.

Philosophers explicitly and economists at least implicitly treat appeals to explanatory power as a *foundational inference rule*. Simply put, the rule works like this: given a specific set of data and two competing hypotheses, infer (or believe or take as better confirmed) that hypothesis that provides the best explanation. To say that this strategy is a foundational inference rule involves three things:

1. The strategy is *primitive*. Other epistemic notions are explicated in terms of explanatory virtues, other inference strategies are justified because they produce the best explanation, and, in general, other epistemic claims get their warrant from use of IBE and not vice versa. Gilbert Harman (1965), to cite just one example, asserts that all inductive inference is really a species of IBE. Lawrence Bonjour (1985) argues that all epistemic justification is a matter of coherence, where coherence is spelled out in terms of IBE. Smart (1989) and Boyd (1985) defend scientific realism on the grounds that it is the best explanation of scientific success. In each case, IBE is taken as primitive, as doing the justifying rather than needing justification itself.
2. The strategy is *formal*. Like modus ponens and other inference rules, IBE makes no substantive assumptions. In other words, it

does not depend on any specific subject matter (hence "formal" as opposed to "material"). This means that the rule holds regardless of what the world is like, of what empirical contingent facts there are, and so on. So IBE has roughly the same status generally attributed to rules of statistical inference, which are thought to follow solely from the laws of probability.

3. The strategy is *sufficient*. Given the data and the competing hypotheses at issue, IBE guarantees that the most explanatory hypothesis is best confirmed. This notion further explicates the idea of being an "inference rule." A valid inference in logic ensures the truth of the conclusion given the truth of the premises; the conclusions of statistical inference are similarly guaranteed, given that the initial assumptions about sampling and so on are satisfied. IBE is supposed to do something similar. In brief, IBEs are indefeasible; they cannot be overridden.

So appeals to explanatory power are supposed to explain and ground other evidential practices; they are to do so without resting on contingent empirical assumptions; and they are supposed to give us inferences that cannot be trumped.

No doubt this description makes IBE fairly powerful. Most defenders and users of IBE are quite unclear on the details of IBE. Yet the virtues they claim for it commit them to something like the three conditions listed above. We can think of them as describing IBE at its most interesting. Weaker versions then are possible but less interesting. IBE might not be *the* fundamental epistemic notion but one among several. IBE might rest on some empirical assumptions, but ones that are minimal and hold across wide domains. IBEs might not guarantee their conclusions but only ascribe them high probability. However, the more such qualifications are added, the less role we are giving to explanation in confirmation.

Having sketched what IBE is supposed to be, I want now to argue that it has no such role; while IBE can be persuasive, it is no foundational inference rule. Rather, I shall argue, IBE depends for its success on substantive, contingent empirical assumptions that are often domain specific; its confirmatory power is only as good as the evidence for the assumptions. Moreover, this means IBE is defeasible. In other words, it is just one sort of empirical virtue and thus can be overridden by other epistemic considerations.

My argument is this: Accounts of explanation fall roughly into two camps, namely, those that equate it with unification and those that make the citing of causes defining. IBE taken as inference to the most unifying theory either turns out to be nothing more than inference to the theory with the most evidence and thus no special inference rule at all, or it presupposes a particular sort of unification, one that will rest on contingent empirical detail and be defeasible. IBE as inference to the best causal explanation will likewise be neither formal nor sufficient, because the best causal explanation will depend on substantive and domain-specific assumptions about causes and because the hypothesis that fits best with our causal knowledge can be inadequate on other grounds.

The argument just sketched turns on no very specific account of explanation. Most, perhaps all, extant views about explanation fall into either the unificationist or causal camp as those terms are meant here. The views of philosophers like Hempel (1965), Friedman (1974), and Kitcher (1989) or economists such as Wilber (1978), for example, are various versions of the idea that explanation comes via unification. For the argument that follows, unification can be taken either as a property of the world or as a property of intentional objects such as beliefs or theories, thus allowing it to be neutral on whether explanation invokes objective properties about the world, subjective ones about ourselves, or some complex combination of the two. The causal camp includes those who think explanation comes via citing mechanisms, fundamental causes, causal structures, causal processes, natures and tendencies, and so on. Both approaches can grant as helpful elaborations various pragmatic approaches to explanation. Moreover, nothing in the argument that follows assumes that all approaches fall into one and only one of these two categories—complex combinations of causal and unifying properties are possible as well.

IBE understood as inference to the most unifying hypothesis is either empty or it is neither formal nor sufficient. It is empty when unification is cashed out in other epistemic terms. Harman (1965), for example, equates the best explanation with what is less ad hoc and more plausible. Advocates of Bayesian confirmation theory often define best explanation directly as the hypothesis best supported by the data (Howson and Urbach 1993). Economists frequently do the same: they claim a given model is superior because it best explains when they clearly mean that it has the greatest predictive scope. These versions of IBE are uninteresting because the notion of explanation is doing no independent work—IBE is nothing more than inference to the hypothesis that best coheres with our overall belief system. In short, IBE is nothing more than the claim that the hypothesis with the greatest total evidence is best confirmed. Few can disagree with IBE in this form, but explanatory strength is doing only rhetorical work.

IBE becomes more interesting when unification is given some specific content beyond mere total coherence. Unification may then be understood as fitting with some special set of beliefs—for example, fundamental laws—or as a particular sort of coherence, for example that which is provided by theories with the fewest independent assumptions. When explanation is cashed out in these ways, IBE is no longer trivial. Now the problem is that appeals to explanatory power are neither formal nor sufficient—they rest on substantive empirical assumptions and can be trumped by other epistemic factors.

We can see why this is by looking at one account of unification in a little more detail. Kitcher (1989) argues that explanation comes from the kind of unification made possible by *argument strategies*. The theory of evolution explains by applying the Darwinian argument pattern across a diverse range of organisms, traits, and environments. That argument pattern involves a general schema describing the forces of natural selection. Explanation comes as we show that a single schema can be applied again and again by filling in local detail. Argument patterns thus unify, and how well they do so is roughly a function of (1) the simplicity of the pattern and (2) its breadth of application. If Kitcher's account is right, IBE then would require us to infer the hypothesis that fits with the most comprehensive argument strategy.

Kitcher's account is insightful and suggestive, and, as I have argued elsewhere, has useful things to say about economics (1996 chpt. 7). However, it surely leaves IBE with something less than a foundational role. Fitting with the Darwinian argument pattern is good evidence for a hypothesis only on two preconditions. First, the substantive facts assumed by that pattern must be well confirmed. An argument pattern whose past applications met with mixed success would be weak grounds for inferences about new cases. Second, we must have good reason to believe that the context under consideration is sufficiently similar to those previous cases where the Darwinian pattern has been applied. For example, if the data to be explained come from a situation where nonselective forces like drift might be significant, then the second condition is not met, and it would be a mistake to infer to the hypothesis most consistent with the Darwinian argument pattern. These requirements thus clearly make IBE dependent on prior substantive empirical assumptions. They also strongly suggest that IBE is defeasible. A hypothesis that fits well with the Darwinian argument strategy but results, for example, in less precise and accurate predictions or that requires ad hoc assumptions would be to that extent suspect. Nothing about IBE prevents such other factors from overwhelming the value of explanatory power.

These general points about IBE are even easier to see if we take explanation as the citing of causes. Inferring to the best causal explanation requires us to use our background beliefs about causation. Those beliefs may range from quite general notions about what causation is to very specific assumptions about what causal variables might be relevant to the data at hand. Clearly, inferences relying on such assumptions will be neither formal nor sufficient. Arguably our basic notion of causation depends on the results of empirical investigation; surely our assumptions about which causal processes are relevant to what data do so as well. Moreover, any IBE based on causal explanation is bound to be defeasible for an obvious reason: the hypothesis that best fits with our background causal knowledge may be a poor hypothesis on other epistemic grounds. Newtonian physics, for example, fared poorly *as an explanation* by the lights of the dominant mechanical philosophy, which required no action at a distance. However, Newton's ability to provide a simple set of laws that predicted the observed phenomena with great accuracy outweighed any explanatory failings. Inferences to the best causal explanation were inferior inferences.

If appeals to explanatory power are not a foundational inference strategy, that does not mean they have no role. However, their proper place has to be reconsidered. The force any IBE has depends on substantive background assumptions. It is those assumptions that are doing the epistemic work, and it is those assumptions that must be evaluated to assess appeals to explanatory power. So simply claiming that one's favored model better explains is not enough. The specific assumptions about explanation lurking behind that claim must be plausible. Consequently, IBEs will generally depend on domain-specific knowledge for their force and to that extent will be context dependent. Moreover, explanatory power, even if real, must be evaluated against other empirical virtues.

What does all this tell us about evaluating neoclassical economics and its individualism? For one, it means that to evaluate the explanatory power of neoclassical economics, we must look at the specific presuppositions neoclassical theory makes about explanation and the economic world. It also means that even if neoclassical economics does win on explanatory grounds, it might nonetheless be inferior overall because of other empirical failures. In what follows I focus on these two possibilities, with primary emphasis on the first. I break my discussion into two parts, using the distinction made above between explanation as unification and explanation as causation.

The Unifying Power of Neoclassical Theory

Perhaps the most powerful-sounding defense of neoclassical theory goes this way: "Neoclassical theory provides us with a mathematically elegant set of postulates that can be applied over and over again to economic phenomena of apparently quite diverse nature and indeed to social processes not normally thought of as economic at all. By showing that these many different processes can be derived from a small set of basic postulates, neoclassical theory shows that it has the kind of explanatory power characteristic of good science. None of its rivals have anything like this ability to unify. Neoclassical theory is clearly far and away superior."

Of course, the reasoning is not always as explicit as this. But versions of this argument are common. This reasoning also lies behind the most common criticisms of alternative approaches, namely, they are *atheoretical*. Economists often value theories not so much because they allow precise empirical predictions and testing, for example, but because they allow us to apply elegant models to diverse phenomena—in short, because they unify.

This defense turns crucially on the idea that explanation comes from unifying and that neoclassical theory does just that. In this section, I evaluate this line of reasoning using the results of the previous section. I raise three fundamental objections to this line of defense: (1) that appeal to unifying power does little to solve the problem of unrealistic assumptions, the most vexing problem for neoclassical economics; (2) that the unifying power of neoclassical theory comes only via substantive assumptions about the economic world that are either empirically highly questionable or that identify variables inconsistent with the neoclassical and individualist programs; and (3) that unifying power itself is a problematic notion of explanatory power.

The most powerful and frequently voiced objection to neoclassical theory comes from the heroic assumptions built into its models. Recall the neoclassical model sketched earlier: it assumed price taking, full rationality, complete markets, and so on. Yet those conditions are seldom met in reality. So the long-standing puzzle is how neoclassical models can explain real economic processes. Neoclassical economists hope to defuse this worry by pointing to explanatory power—to the fact that their models can be applied over and over again to diverse phenomena.

That hope is misguided. On Kitcher's account we unify by showing that a fundamental argument pattern applies to diverse phenomena.

That requires showing that the pattern *applies*—that the basic factors it cites are present. For example, in the Darwinian pattern we must show that there are the relevant traits, genes, and inheritance. If we know that in reality none of these factors are really present, then we cannot claim the pattern applies and thus cannot claim to unify. However, neoclassical models frequently rest on assumptions we know to be false.

When those models are applied to a diverse range of phenomena, we thus have, at most, potential explanations—explanations of how the phenomena might happen. But it is a shaky move indeed to infer from "*H* potentially explains *E*" that *H* is true. The hypothesis that we are under the control of extraterrestrial beings *potentially* explains much, but its presuppositions are false.

Of course, there are degrees of irrealism, and not every simplification or idealization undercuts IBEs. In the next section, I shall say something about how to factor such devices into IBE taken as inference to the best causal explanation. However, I know of no worked out way to incorporate simplifications, idealizations, and the like into the unificationist account of explanation. Roughly put, however, we want to know that the model is "close enough" to reality to explain. Neoclassical theory describes an entity we might call the "Walrasian spot market" where consumers and firms have full information, there are no barriers to entry, and so on. So we need to know whether any given application of a neoclassical model is close enough to the Walrasian spot market to make explanation possible. That is a substantive question in economics that requires examining the particulars of the economic situation in each case. Simply appealing to overall explanatory power is not enough. Thus solving the irrealism problem forces us to move beyond IBE to contingent factual information about economic processes, my main moral.

If we do look at the empirical assumptions needed for neoclassical theory to unify, I think we will find that they are either empirically dubious or ultimately inconsistent with the neoclassical, individualist program. Of course, no general proof of these claims is possible; a convincing case would require a careful look at representative attempts to extend neoclassical theory to diverse phenomena. Here I want to make a first step in that direction by looking at new work on the neoclassical theory of the firm. It shows, I shall argue, that neoclassical theory unifies only by making some questionable empirical assumptions.

Standard neoclassical theory until recently had no theory of the firm. Or, put differently, the theory it had said nothing about the internal workings of the firm. Firms were treated as if they were single individuals hiring inputs that were then combined according to a technological production function, given prevailing market prices. Recently, however, neoclassical theory has tried to do much more—to apply the maximizing under constraints model to the individuals *inside* the firm rather than acting as if the firm itself were an individual. This means taking on the organizational and social processes that make up the firm, something traditionally a domain for sociologists or for economists outside the neoclassical tradition such as institutionalists and Marxists. This new work on the firm is thus a prime example of neoclassical economics' individualism and of its apparent explanatory power.

The traditional neoclassical theory of the firm takes firms to be profit maximizers. Yet neoclassical theory also takes workers, managers, and the like to maximize their own self-interest. However, if the owners of a firm are not its managers—as is often the case—then the profit maximizing firm would seem to be at risk. Much interesting new work on the firm asks when and where these two postulates—profit maximizing firms and self-interested managers—are compatible. I have in mind particularly the work of Fama (1980), who has been an innovator in extending neoclassical account inside the firm while arguing that profit maximizing remains a plausible claim about firms as a whole. His basic argument is this: The known data about firms—that stockholders and managers are distinct, that salaries rise as managers have greater authority, and so on—does not refute but in fact supports the traditional hypothesis of a profit-maximizing efficient firm. The implicit simplification that risk takers and bearers are identical is false but not damaging. If we factor in the complications that come from a more realistic account, we will find that the profit maximizing efficient firm is consistent with the data about ownership and control, salaries, and so on. The traditional causal model—that market competition produces profit-maximizing firms that equate marginal cost and marginal revenue—is still well confirmed, because our best understanding of the complicating causes suggests that it is profit maximizing that best explains the data. In short, it is the main explanatory variable, not the simplifications, that predict the data.

Fama argues for this conclusion by citing four causal processes: the competition between firms whose managers and owners are identical and firms where the functions are separate, the market for managerial talent, the market for outside directors, and the effects of potential takeovers. These processes give a more realistic description of firms and their environments. Each shows that the data on ownership, control, and salaries in fact support the maximization hypothesis. A manager's price in the market for managerial talent will be determined by his or her contribution to firm success. Managers who shirk will be found out when they change jobs. Moreover, the contributions of managers to firm success will depend on the efforts of managers below and above them, thus giving them strong incentives to monitor their behavior. Similarly, the value of an outside director will depend on the contribution he or she makes to firm profitability. The threat of outside takeovers has a similar effect—forcing managers who are not owners to maximize firm profits. Finally, in a competitive market, firms with managers who do not maximize profits will be eliminated in competition will those who do. Thus the separation between owners and managers does not entail that managers can ignore profitability and efficiency.

So there is a market for high-level management that ensures that profit maximization and the pursuit of individual self-interest are compatible. This general approach—treating internal structure of firms as resulting from contracts between self-interested parties—is extended in interesting ways by Rosen (1982) and Williamson (1975, 1985) to explain other data about the firm. Such data include the fact that firms have a hierarchy of authority and of pay differentials, that much hiring is from within, that employment relationships are generally long term, and so on. These facts are also initially puzzling for the neoclassical tradition, since they seem more characteristic of bureaucracies than of fully efficient producers in a Walrasian spot market.

Williamson argues that a firm based on hierarchical authority relations will out compete the other possible ways corporations might be organized. That is because a firm with a hierarchical command structure will minimize transaction costs. Transaction costs are the costs involved in bargaining and enforcing contracts. Traditional neoclassical models have ignored such costs by assuming full information and compliance. Williamson argues that many transactions involve both uncertainty and asset specificity. Uncertainty comes from difficulties in measurement—in determining, for example, the true costs or quality of a product. Asset specificity occurs when one or both parties to a trade must make an investment that cannot be recouped if the trade falls through—as when, for example, a corporation invests in employee training.

According to Williamson, the hierarchical firm exists because it handles these transaction costs. Long-term contracts and internal labor markets solve the problems of asset specificity; they also remove the uncertainty that comes from buying services in an external spot market, for by bringing the service under the control of the corporation, greater information and control are possible concerning the quality of the product, its true costs, and so on. So the internal structure of the firm can be explained as resulting from market processes, once the full costs are specified.

Rosen adds a further dimension to these two accounts. Williamson explains why hierarchy exists and Fama why it is compatible with overall profit maximizing. Rosen in turn explains why wages rise with authority. If we suppose that individuals differ in their managerial talent and that the more managerial talent a supervisor has, the more productive his or her employees are, then differences in salary will follow differences in authority. Rosen is in effect using the classical comparative advantage argument for the deployment of resources. If we ask what use of resources will maximize output at equilibrium, we will find, he argues, that most talented managers will be assigned to the positions of greatest authority in the largest corporations and will receive by far the highest compensation.

These three accounts illustrate the apparent explanatory power of the neoclassical paradigm. A social phenomenon of some complexity—the internal structure of the corporation—is shown to be both consistent with the neoclassical postulate of profit maximizing and to be explainable by the constrained maximizing of individuals in a market. Here seems to be a powerful case for neoclassical theory and the individualism it exemplifies.

These appearances are, however, deceptive. These three attempts to apply neoclassical theory inside the firm rest on specific and substantive economic assumptions for their explanatory power, as we would expect from our account of IBE. Explanatory power on the unificationist view comes from extending an argument pattern across domains. Making that extension presupposes that the new domain is sufficiently similar to the old for the argument pattern to apply. As I suggested above, the natural range of the neoclassical approach is the spot market characterized by many buyers and sellers with fully defined preferences and the other traits of perfect competition. Applying the neoclassical argument strategy to the internal workings of corporations presupposes the substantive claim that those workings are sufficiently similar to the spot market paradigm to make the extension explanatory. That similarity, I would suggest, is not forthcoming. In particular, at least the following interrelated differences seem crucial:

1. *Managerial talents are not easily observable.* In the standard Walrasian market, prices and qualities of goods are perfectly known.

Managerial talent, however, is generally exercised inside a firm and thus is at best indirectly observable by those outside. This raised real doubts for Fama and Rosen's accounts, which depend crucially on an efficient market for managerial talent. Williamson recognizes that goods like managerial talent are often hard to measure, but then assumes that *inside* the firm these problems are eliminated. But in large hierarchies that is not obvious either, especially given problems noted below.

2. *Is managerial talent an individual or joint product?* If and when managerial talent is the result of team production, then the contribution of each individual may not be well defined and certainly makes it harder to observe. This obviously contributes to the previous problems and makes an efficient market for managerial talent still more unlikely.
3. *Are the phenomena to be explained the result of competitive processes at one or many levels?* Fama, Williamson, and Rosen try to explain corporate phenomena as the efficient outcome of a competitive process in strict analogy with a competitive market. However, they mix together processes operating at different levels—at the very least at the level of individuals and the level of firms. Nothing ensures that these two processes push in the same direction. Managers may do well because they are in good corporations, not because they are good managers. Similarly, traits prized by the managerial market may not be traits that promote corporate survival. Managers have a natural incentive to exaggerate their contribution, to claim credit for the work of others, to advance the interest of their unit at the expense of others in the corporation, to focus on short-term profitability at the expense of long-term profitability if they expect to change jobs frequently, and so on. (This latter prospect is especially likely for the processes Fama describes. Shirking managers are found out when they change jobs—that may equate their marginal revenue and marginal product, but it does no good for the firm they used to work for, which presumably was paying them above their contribution.)
4. *Do the relevant consumers have a well-defined preference ordering?* Both Williamson and Fama treat owners of the corporation—the consumers of managerial talent—as if they were a single individual. Yet the empirical evidence suggests that stockholders are, of course, numerous and, more important, have diverse interests and preferences. That means constructing any aggregate preference function will be subject to all the difficulties Arrow

identified. As a result, transferring traditional neoclassical supply and demand analysis to the "managerial market" becomes troublesome.

5. *Are the markets in question ones where assumptions of equilibrium and optimal outcomes are reasonable?* All the work described above assumes that market processes will pick optimal outcomes. The explanatory stories they tell presuppose that competition and rational choice suffice to ensure that, at equilibrium, resources are optimally employed, given the relative constraints. That presumption may hold of the Walrasian spot market. But it is a dubious assumption about managerial markets, internal labor markets, and the like for several reasons: (1) competitive processes act at multiple levels and in different directions as we saw above, making the optimal both unlikely and conceptually troublesome; (2) selective processes may not "see" the relevant factors, because, for example, it is difficult and costly to observe a manager's contribution to a large team effort; and (3) the potential costs and benefits that would go into any assessment of optimality go far beyond those included in the models discussed above. For example, those models all ignore the costs associated with carrying out commands inside a corporation. Managers have to expend resources to get their wishes implemented; lower-level managers expend resources to disguise their true contribution and to influence the decisions of those above them. These costs—ones associated with the kinds of social processes institutionalists emphasize—are ignored by Williamson, Rosen, and Fama. They make it dubious at best that neoclassical models are easily extended to corporate behavior.
6. *Are individuals inside the firm primarily motivated by pecuniary gain?* The neoclassical model is at its cleanest when producers are motivated solely by monetary rewards. Once we allow personal loyalties, prestige, ethnic identities, and other real human motivations into the picture, the elegant results about the distribution of resources no longer hold. It is, of course, possible to complicate the simple models to allow producers to maximize the production of both monetary and nonmonetary goods. But doing so requires at least a clear account of what those nonmonetary goods are. The models discussed above make no effort to include nonmonetary goods, though arguably they are much more important in the social life of the corporation than they are in an impersonal Walrasian market.

So the apparent unifying power of neoclassical theory rests on numerous substantive assumptions about the economic world. Those assumptions are, however, empirically suspect and thus its explanatory strength is correspondingly suspect.

Let me turn now to another doubt about the unifying power of neoclassical theory: what it takes as given. Institutionalists and other critics of the neoclassical tradition often complain that it takes preferences, norms, property rights, corporate culture, and the like for granted. This claim—no doubt true—may or may not be grounds for worry, depending on what conclusion we are arguing for. *If* these factors really are more or less exogenous—uninfluenced by the factors discussed in neoclassical models—then it is surely a mistake to conclude that neoclassical theory is inherently flawed because it does not explain them. After all, studying partial systems by taking background causes as fixed is common practice in our best science.

However, if neoclassical theory claims to fully explain, to better explain, or to show that social processes can be explained individualistically, then there are indeed problems. To the extent that neoclassical theory takes institutional structure as given, it has not eliminated social explanations but presupposed them and thus cannot be a full competitor to social explanations. To the extent that neoclassical theory unifies by assuming a set of background practices, its unifying power is bought on the cheap. Neoclassicals frequently charge that alternative approaches are piecemeal and atheoretical. Yet if the breadth of the neoclassical approach comes simply from ignoring many economic phenomena, then its ability to unify counts for much less.

My final doubts concern the role of unification in explanation and confirmation—I doubt that unification is central to explanation and, following the arguments of the second section, I doubt that the theory that best explains is best confirmed. My worries about unification are multiple. (1) There are often multiple ways to unify any given set of beliefs, raising the specter that unification is subjective. (2) It is unclear how we measure the complexity of an argument pattern or its scope and how we trade off those two factors against each other. (3) Unification often is important not because of its role in explanation but because of its importance in testing for consistency, its heuristic role, and so on; unifying argument patterns often look like good strategies for finding the causes, where the causes are doing the explaining. (4) Much good science seems bereft of general argument patterns; molecular biology, we saw in chapter 4, is a good example here. The moral I draw from these difficulties is that we do best to focus on causation as central to explanation rather than unification. Of course, defenders of the unificationist approach may eventually resolve these issues. Until they do, however, we should be weary of putting too much stock in the ability to unify.

Finally, I doubt that neoclassical theory would be well confirmed even if unification were central to explanation and neoclassical theory best unified. A theory that meets explanatory standards can fail on other empirical grounds (recall Newton and action at a distance), with those grounds outweighing the explanatory successes. I think this may well be the case with neoclassical economics, though sustaining that claim would take a much more detailed look at actual empirical work than is possible here. It may be that an eloquent neoclassical story can be concocted for quite diverse phenomena. Yet we can likewise derive numerous implications from neoclassical theory that fail to fit the facts. Some of those implications concern the substantive assumptions needed to generalize the neoclassical model from simple spot markets to more complex social phenomena. Others involve sundry, unconfirmed implications like the absence of discrimination in competitive markets. Of course, these failings may all be excusable due to the holism of testing. Yet it will surely not do to ignore these other empirical difficulties simply on the grounds that neoclassical economics has great explanatory power.

Is Neoclassical Theory Inference to the Best Cause?

So far I have argued that evaluating theories by their explanatory power depends on substantive empirical assumptions, that explanatory power can be trumped by other empirical virtues, that neoclassical theory gains its ability to unify only by making numerous doubtful empirical presuppositions, and that therefore explanatory power—taken as unification—will not save neoclassical theory and the individualism it embodies. However, explanation is often explicated as the citing of causes, and I have said nothing directly about neoclassical theory's ability to provide the best causal explanation. It is to that topic I turn in this final section. My general thesis is again that defending neoclassical economics as an inference to the best cause rests on substantive and largely dubious economic assumptions.

A standard rationale for the neoclassical program goes like this: We know that individuals are in the large self-interested. We also know that they learn from errors and thus are largely rational in pursuing their goals. Moreover, we know that competitive market mechanisms reinforce these behaviors. In addition, we know that every economic action involves constraints—the use of a resource in one way inevitably means forgoing alternative uses. These are the fundamental causal factors in the economic realm. So any adequate theory must be consistent with them. But neoclassical models not only describe such causes but do so in a systematic way; alternative approaches do not. Neoclassical theory is thus superior because it provides the best causal explanation.

This reasoning looks convincing at first glance. But it faces a variety of problems. Let's begin with the realism of assumptions problem—the fact that neoclassical models often make heroic ceteris paribus assumptions. As we saw in the last section, unificationist approaches to explanation have trouble analyzing such provisos. Causal approaches have a more natural way to handle unrealistic assumptions, but they do so again only with the aid of some quite specific empirical knowledge. They must provide us with sufficient causal information to show that the model in question would produce the data at hand because of its main variables, and not simply because of the literally false assumptions it makes. How can we have such assurances? There are numerous ways to achieve this confidence, and some of them can be seen as inferences to the best causal explanation. For example, a model resting on unrealistic assumptions can nonetheless in part be confirmed if

1. We know just what the potential complicating and confounding factors are and can show (or produce) circumstances where they are in fact not present and that the model successfully predicts in those circumstances.
2. We know what the potentially complicating and confounding factors are and can show that even when present they would have little effect.
3. We know what the potentially complicating and confounding factors are and can show how adding their influence to our model produces more accurate predictions.

Methods like these help ensure that it is our explanatory variables, not the simplifications we make, that are responsible for predictive success. Each is a kind of inference to the best causal explanation. Given the data and what we know about causes, we infer that it is our hypothesized variables, not the simplifications of our model, that produced the data.

It is important to note that we need such specific causal information to defend unrealistic models rather than some more nebulous fit with our general understanding of causes. If, for example, we believe that all economic outcomes result from self-interested maximization under constraints, fit with that information is not sufficient to confirm an unrealistic model—not, at least, if our concern is its irrealism. Fit with a general causal picture of the world does not by itself show that our model is predicting the data legitimately; we need instead the much more detailed information that allows us to infer in the specific case at hand that simplifications and idealizations are not confounding. So any causal IBE is only as good as the causal knowledge on which it is based and it is that knowledge that is doing the real work. If we have little information about *what* are the potentially confounding factors or *how* they interact with our hypothesis, then an IBE defense of an unrealistic model will be correspondingly less persuasive. So the irrealism of assumptions problem cannot be solved by quick appeal to inference to the best cause.

This moral is reinforced by the fact that neoclassical postulates about causes are often compatible with quite different models of the same concrete phenomena. These differences come in part from different kinds of simplifying assumptions made or the different ways neoclassical postulates can be interpreted—as, for example, in different specifications of what rationality involves. So fit with the neoclassical vision of self-interested agents acting under constraints by itself will not tell us which simplifications are justified and which are not.

Even when unrealistic models are defended by substantive assumptions about causation, they may nonetheless not be the models that are, all things considered, best confirmed. Take two models M_1 and M_2, where both rest on (different) simplifications and where M_1 is the best explanation in the sense discussed above. M_2 might still be more reasonable to believe because of its other epistemic virtues. If M_2 more accurately predicts the data and we believe that the causal knowledge favoring M_1 is relatively weak, then M_2 might be the best confirmed. Sometimes our understanding of the complicating causes is relatively poor. Thus, even though we believe one model to have ruled out complications more successfully than another, relatively poor predictions on its part may reasonably lead us to reject it for a competitor that better predicts despite its troubling simplifications. In short, IBE defenses of unrealistic models are defeasible.

The upshot is that a quick appeal to causal explanatory power will not do—solving the irrealism problem, like good IBEs in general, requires that the empirical presuppositions about causal explanation are well confirmed.

I want to turn now to look at some of the actual empirical assumptions involved in the neoclassical account of "good causal explanation." Neoclassical theory makes at least two crucial assumptions about good explanations: that any outcome must be consistent with self-interested behavior and that economic institutions and behaviors, so long as they result from a competitive economic process, exist because they are optimal. Both claims, I want to argue, rest on contingent, substantive, and empirically dubious presuppositions about the economic world. Thus, neither strongly supports the neoclassical program and its individualism.

Let me begin with the idea that the best causal explanation is one consistent with self-interested behavior. One way to see that this requirement makes quite substantive assumptions is to note that it does not follow from or embody any general methodological rule of science. Neoclassicals sometimes defend the self-interest postulate on the grounds that every scientific explanation must cite a mechanism, and the mechanism behind economic processes is self-interest. Yet the demand for mechanisms, as we saw in chapter 2, is not a requirement for good science in general nor would it necessarily support the neoclassical approach over others.

Mechanisms may not be essential for several reasons: (1) "the" mechanism is of dubious sense, for we can generally find mechanisms at diverse levels; (2) we can often confirm theories without any particular account of underlying processes, especially when higher-level accounts are well confirmed and make no very specific assumptions about underlying causes; and (3) innumerable causal explanations succeed without providing underlying mechanisms—we explain that balls break windows and planets attract without being able to tell the relevant quantum mechanical story. So the neoclassical emphasis on self-interest does not rest on a general principle of scientific methodology.

Moreover, even if mechanisms were essential, it would take further assumptions to show that those mechanisms must be in terms of individual self-interest. First, there is the question why mechanisms must be at the individual level—after all, even neoclassical economics has long provided mechanisms at the level of corporations and households, not individual human beings. Thus, evolutionary and institutionalist approaches might well meet the demand for mechanisms, but do so by describing processes not at the individual level; Nelson and Winter's (1982) version of evolutionary processes works exactly this way. Furthermore, even if we wanted mechanisms in individualist terms, it would still be an open question whether those mechanisms appealed to self-interest rather than some other set of motives. Thus requiring mechanisms does not by itself require the self-interest postulate.

We can also see that the self-interest postulate depends on highly substantive assumptions by simply asking what it requires. The neoclassical tradition vacillates between what we might call "thin" and "thick" versions of the self-interest postulate, where the thicker the version, the more substantive assumptions made. We can break the self-interest postulate into claims about processes and about ends or goals. Thin notions put very few constraints on goals. Typically they require pursuing preferences, where the preferences can range over any set of objects—they can be other directed for instance—so long as they exhibit the coherence described by transitivity and the like. Alternatively, thick notions require quite specific goals; the most substantive comes from equating self-interest with the pursuit solely of financial gain. Thick notions likewise require much of processes. Individuals are claimed to maximize and to do so subject to relatively few constraints. Initial endowments may be set, but thick notions of the self-interest assume great cognitive powers and few or any costs to their use—in short, extensive knowledge of market and production variables. Thinner notions allow satisficing or even weaker search heuristics and factor in the costs of information gathering as well.

I point out these differences in some detail because they raise a fundamental dilemma for making the self-interest postulate essential to good explanation. The dilemma is this: thick notions require highly substantive and specific assumptions that tend to be implausible as constraints on good explanation in the social sciences; thin notions are much more plausible, but do not favor neoclassical approaches over other alternatives. If preferences can be entirely other directed, nonpecuniary, and pursued by use of heuristics and routines, then the self-interest postulate is consistent with most alternatives to the neoclassical approach. Yet the thick versions of the self-interest postulate are notoriously at odds with the data. So where the constraint is plausible, it does not distinguish between approaches; where it does so distinguish, it is implausible.

I want to illustrate this dilemma by looking briefly at the controversy over rational expectations and the need for microfoundations in macroeconomics. Defenders of the New Classical approach have argued that their theories are better because of their greater explanatory power. Keynesian macroeconomics, they claim, is essentially ad hoc in the causal mechanisms it invokes. Only macroeconomic models that are grounded in rational individual behavior can provide adequate explanations. Thus, models with rational expectations are always superior to those invoking money illusion, price stickiness, animal spirits, and the like. New Classical theory is more plausible because it provides better explanations.

It is not hard to see that these claims rest on substantive assumptions about explanation. The presumption that any economic explanation must result from rational behavior is one such assumption, but that really is only the tip of the iceberg. There are both deeper and more restrictive assumptions at work. The deeper assumption is that any macrolevel explanation requires microlevel causal detail, a principle rational expectations advocates explicitly affirm (Begg 1982, 5). The more restrictive assumption is that mechanisms based on rational behavior must be those described by the New Classicals. Typically that means individuals have expectations that reflect the true model of the economy and that all potential gains from trade have been eliminated (i.e., that markets clear).

Surely these assumptions presuppose some very specific assumptions about explanation. The general claim that every macrolevel explanation requires micromechanisms is highly restrictive. As I suggested above, explanations both scientific and mundane seem to violate it: biologists often explain traits by their selective value while remaining ignorant of the underlying molecular details and errant baseballs explain broken windows despite our ignorance of the quantum mechanical details. So the demand for mechanisms is not some universal truth about good explanations; it is a contingent and substantive assumption. Equally contingent is the implicit assumption that the rational mechanism must be a rational expectations process in a clearing market. As the critical literature has shown (see, for example, Hoover 1990; Hillier 1991; Peel 1989; Sent 1997), this equation is not inevitable. Expectations might be rational in at least three senses: (1) they might correspond to the true or correct model of the economy; (2) they might make complete use of all available information; and (3) they might make use of as much information with as much efficiency as is optimal given the costs of collecting and processing that information compared to the gains from doing so. New classical models typically have invoked either (1) or (2) to get their results. Yet that requires a quite specific and controversial notion of rationality, for (3) is surely an equally plausible competing notion. Furthermore, market clearing—which has been shown to be essential to the policy neutrality thesis—is obviously an add-on to the idea that individuals act rationally. So the new classical claim to provide better explanations rests on some very specific presuppositions indeed.

Sent (1997) has traced out such presuppositions in great detail in Sargent's work on rational expectations. For Sargent, rational expectations meant a notion of rationality that (1) left no place for data gathering, since rationality is equated with using traditional econometric practices on preexisting data, (2) assumed the current mishmash of rationales for statistical practice are the correct ones, despite the unresolved disputes between classicals and Bayesians, (3) attributed even greater rationality to agents than to econometricians by allowing the latter, but not the former, to have error terms in their estimations, and (4) entailed that agents have no reason to trade (see Tirole 1982). These presuppositions are, of course, strongly substantive. They show that the New Classicals' claim to best explain depends on a highly specific and contentious interpretation of the rationality postulate.

The presuppositions described above are controversial, but let's assume that they were not. Would that mean the New Classical approach was the most reasonable to believe? It is not at all obvious that it would. The reason is that explanatory power is only one virtue. Macroeconomic models embodying the rational expectations assumptions are apparently predictively inferior, regardless of their explanatory virtues. Eckstein's (1983) work nicely illustrates how this can be. He tested the DRI large-scale macroeconomic model against various data with and without rational expectations assumptions. When equations were modified to reduce or eliminate the prospect of money illusion by workers, the predictive error grew by a factor of four in tests on historical data. Investment equations that allow for the possibility of systematic error predict the data better than those in which error is ruled out. Finally, the DRI model without strong rational expectations assumptions does not predict poorly for data gathered after shifts in government monetary policy, as it should on the rational expectations assumption. So Eckstein's evidence—also supported by Fair's work (1984)—shows that however appealing their explanatory assumptions may be, rational expectations models are apparently inferior predictors.

I want to turn now to another neoclassical constraint on good causal explanation and argue that it is likewise contingent, substantive, and empirically dubious. The constraint I have in mind claims that good causal explanations must show that the explanadum resulted from an optimizing process. Neoclassical theory is built on the idea that the best explanation must show how the phenomena in question can arise as the optimal outcome of maximizing under constraints. For example, consider the recent work on the theory of the firm mentioned earlier. There are many facts about firms that initially seem at odds with the neoclassical outlook: long-term contracts, internal promotion, and wages and salaries apparently in excess of marginal products. Neoclassical theory does not see these facts as a call for sociological explanation in terms of norms, power, and the like. Instead, an adequate explanation shows how long-term contracts maximize the returns to individuals facing constraints on information, for example. Wages in excess of market rate are "efficiency wages" that contain an additional payment in return for not shirking. Internal promotion occurs because of asset specificity, informational asymmetries, and the like. In short, whatever explains these facts about the firms, they must result from some kind of optimizing process. This sort of reasoning is endemic in the neoclassical tradition.

As with the self-interest constraint, we can give either a thin or thick description of "optimizing processes." The thicker the account, the more clearly it rules out competitors to the neoclassical approach *and* the more substantive and empirically dubious it is. Below I focus on the problems for a thick account—problems for requiring optimizing processes like those found in the ideal spot market. At least the following obstacles make it unlikely that real-world processes are optimizing:

1. *The requisite markets often do not exist.* (a) Goods supplied at different times are importantly different in their economic consequences. So optimizing requires markets for goods delivered at different times in the future. However, most market economies do not have extensive futures markets, making optimal outcomes unlikely. (b) If information is costly, then markets become fragmented—search for price and quality information is limited and thus many sellers will not guarantee a competitive market exists. (c) Barriers to entry are common and take on many subtle forms. When there is learning by doing, then start-up costs for new enterprises may be prohibitive, even if there are in theory monopoly rents to be appropriated by new entrants. Strategic behavior—for example, predatory pricing—can likewise make the establishment of new firms impossible.
2. *Information is costly and unevenly distributed.* Unlike the ideal spot market, information about prices, quality, technological relationships, and so on is gained only with time and resources. Different players in the economic processes will also have differential access to information simply because of their economic position—as, for example, individuals know more about their traits than do employers, and current employers know more than potential employers.

3. *Competition occurs at multiple levels*. As we saw in the third section, competition between firms and between individuals need not pull in the same direction. The simple spot market has only one level of competition, but once multiple levels are admitted, "the" optimal becomes ambiguous.
4. *The outcomes of real competitive processes depend on historical contingencies*. The initial distribution of resources can strongly affect incentives. Concentration of ownership provides different incentives than distributed ownership, as we saw earlier in discussing managers and the firm. Firms already in a market may have insurmountable leads over firms just entering, making outcomes path-dependent. Local optimums may make it impossible to reach the overall optimal, even when competitive processes are strong. And even when optimizing processes are present, they take time—and thus any given state of the economic world may be a disequilibrium state and thus not at the optimum.

The upshot: requiring that good causal explanations identify optimizing processes rests on the dubious assumption that optimizing processes dominate the economic world. So neoclassical theory cannot claim to explain better than institutionalist, Marxist, and other more social approaches simply because they do not describe the optimal. Of course, I have attributed a thick notion of optimizing processes to the neoclassical tradition. If we drain the notion of any very specific content, then we might add in all the above complications—by allowing for local optimums, optimums given information constraints, optimums relative to initial distributions, and facts about path dependent processes, and so on. But then standard neoclassical results will not hold, and the resulting explanations will look much more congenial to alternative approaches. The claim to better explain will evaporate.

I want to finish by pointing out some ways my argument is incomplete and by making clear what it does not show. I have been arguing that neoclassical theory may not best explain because it rests on dubious empirical assumptions and thus to that extent it provides little support for the individualist program. But since explanatory power is a relative comparison, this is only part of the story. No matter how weak neoclassical theory may be, its competitors may be even more suspect. And neoclassical economists are indeed suspicious of the typical sociological explanations often offered by institutionalists, Marxists, and the like. I suspect that their doubts are only partially justified, but nailing down my argument would require showing that in detail, something I cannot do here. Note, however, that this "we are not as bad as our competitors" strategy would not establish very much—in fact, it reinforces my claim that relative explanatory power is only one of many empirical virtues. It would be odd indeed to take a theory as well confirmed because of its relative explanatory power *if* it were only the best of some very poor explanations.

A second, important caveat concerns the scope of my conclusions. I have not shown the neoclassical approach worthless, inherently inferior, or any such thing. I have rather raised doubts about one standard rationale for that program and thus indirectly for the individualism it embodies. Vague appeals to explanatory power will not do—economists have to get down to the hard work of testing their models case by case against data, doing their best to rule out competing explanations and confounding factors, especially those introduced by simplifications and idealizations.

7

Individualism and Rationality

Individualism seems most compelling when it comes to rationality. It is, after all, individuals who decide what to believe and how to act. Insofar as they do so rationally, the roots of their action and belief seem fully explicable and only explicable in individualist terms. This chapter challenges this deeply held doctrine, one common to much of the Western intellectual tradition. My targets are individualist assumptions in the theory of rationality in epistemology and philosophy of science and in the theory of rational choice in economics. I make the same claim in both cases: rationality cannot be fully grasped in purely individualist terms.

One common theme in what follows is that individualist notions of rationality look much less compelling if we take seriously the idea that there is no first philosophy, an idea I shall assume rather than argue for. Traditional conceptions of rationality are of a piece with the foundationalist project of finding a priori conceptual truths about justification and inference. This tradition also holds that agents are rational only to the extent that they reason in accordance with such truths. Though this program had an official death with Quine, who argued that there is no domain of conceptual truths immune from empirical considerations, few have traced out its implications for individualism and rationality. I will argue below that these implications include the following:

1. If we reject the idea of a priori, purely conceptual truths about epistemic norms, the natural alternative is to see epistemology as an empirical discipline. If epistemology is ultimately empirical, then we should be wary of alleged conceptual arguments showing that rationality must be an entirely individual affair. Moreover, an empirical approach to epistemology asks how our beliefs are

formed and whether those processes promote our epistemic ends. Yet social processes, institutions, and so on are arguably important causes of our beliefs. Thus the demise of foundationalism opens the way to a more social conception of rationality.

2. If we reject the idea that epistemic notions can be fully described by purely formal rules, then any epistemic assessment must rest in part on substantive information about the domain in question—that was a moral of the previous chapter. However, this conclusion bodes ill for the idea that we can describe what rational actors do simply by working out the postulates of rationality. Since individualist notions of rationality make precisely this assumption, the way is again opened for more social conceptions.
3. Finally, the demise of foundationalism suggests that we need to evaluate rationality in practice—according to the situations and constraints agents actually find themselves rather than according to impossible demands of the ideal reasoner. So rational behavior and belief is *responsible* behavior and belief, given the agents' actual situation. But real agents are in situations where social factors play an important role, and they surely face limitations in resources and information. This means that even if we had an a priori logic of justification, it would not tell us much about what real agents ought to do to be rational.

In short, traditional individualist notions of rationality depend on foundationalist assumptions; rejecting those assumptions raises serious questions about the traditional notions of rationality.

Of course, these arguments are sketchy, but they will be fleshed out as we proceed. The chapter works as follows: the next section takes up individualist assumptions in epistemology. I argue that rationality is irreducibly social for three reasons: (1) there can be rational communities whose behavior is brought about by diverse, often nonrational behavior at the individual level, (2) that a group of rational individuals need not be a rational community, and (3) that adequate accounts of individual rationality implicitly invoke the social—they are not genuinely individualist at all. As may be obvious, these are versions of the three problems that usually confront reduction, namely, multiple realizations, context sensitivity, and presupposing social facts. The third section then turns to individualist assumptions in the theory of rational choice, focusing on rational choice in economics—particularly on human capital theory, game theory, and rational expectations. Once again I argue that rational-choice explanations cannot fully capture collective phenomena and that, stripped of social assumptions, they cannot fully explain individual behavior either.

Several qualifications of my conclusions are in order. First, my arguments remain empirical; I do not pretend to show on a priori grounds that rationality must be a social phenomenon. If rationality is social, it is so for contingent, empirical reasons. Second, the arguments of this chapter are preliminary. That should not be surprising. I discuss fundamental issues in epistemology and rational-choice theory; I argue for quite strong claims. My goals are to sketch a plausible prima facie case.

Individualism and Epistemology

Individualism as an epistemic thesis claims that we can fully explain, capture, describe, justify, and so on our epistemic practices in individualist terms. We can in turn take this claim in at least two ways, depending on whether we want individual practices described in epistemic or nonepistemic terms. "The social or collective aspects of rationality are reducible to some complex sum of individually rational actions" is an example of an epistemic version; "the social or collective aspects of rationality is reducible to individual behavior" with no stipulation that the behavior be rational illustrates the alternative version. For example, the first claim would hold that "being a rational research group" is just equivalent to some collection of individually rational actions. The second would hold only that "being a rational research group" is equivalent to some set of individual actions by members of that group, actions themselves that might be nonrational themselves—as would be the case if habitual adherence to authority produced collectively rational outcomes. However, individualists typically defend the first, logically stronger claim that rational communities are nothing but collections of rational individuals. Thus, that will be my prime target in what follows.

Epistemic individualism, like reductionist programs in general, faces at least three potential obstacles: (1) epistemic phenomena at the group level may be multiply realized in individual behavior, thus thwarting reduction; (2) the collective significance of individual behavior may be context sensitive, showing that the facts about individuals may not fix or determine the epistemic facts about groups, thus violating another requirement for reduction; and (3) accounts of individual epistemic states may make essential reference to collective or social phenomena, showing that individual accounts presuppose social facts rather than reduce them. In what follows I make a prima facie case that all three obstacles are real. Before doing that, however, I turn first to clarify the claim that rationality is a social or collective property and to defuse alleged conceptual facts that would show such a notion incoherent.

The idea that rationality is social involves several different theses. One central claim is that we can have good reason to assign epistemic traits to groups and that those ascriptions cannot be reduced to the epistemic traits of individuals, or more strongly, that they cannot be reduced to individual behavior at all, epistemic or not. The second main claim concerns individuals: ascribing epistemic traits to individuals requires reference to social factors, or more strongly, to the epistemic traits of groups. This claim about individuals can also be evidence for the claim that group traits are irreducible—it is obstacle (3) listed above—but the claim about groups can be true even if the claim about individuals is false.

It is important to note that the idea of group rationality advocated here goes considerably beyond the notion of "social epistemology" defended by reliabilists like Goldman (1986) and Kitcher (1993). Both agree that social factors play an important role in rationality and justification. Yet for both the question is whether social factors produce rational *individuals*—individuals with justified beliefs, true beliefs, and so on. So in the end the locus of epistemic evaluation is still the individual, and community rationality is derivative.

The view I am defending claims that communities can be rational in a full-bodied sense.[1] In other words, if community practices efficiently meet epistemic ends, the community is rational *regardless of whether any one individual has the full truth or the full justification.* A body of information can be rationally held, produced by reliable community structures, justified, and the like for the community as a whole.[2] I begin my defense of this claim with conceptual or metaphysical objections. Community rationality as defended here might seem unacceptably ontologically promiscuous. If I claim that communities are rational, justified, have knowledge, and so on *sui generis*, then aren't I committed to group minds and the like? I do not think so, and we can use the general whole–part relation defended in earlier chapters to see why. Communities are exhausted by and supervenient upon the behavior, beliefs, and practices of individuals. In the end, all social structure is a concatenation of these basic entities; collective beliefs do not exist as independent entities. Group belief is thus ultimately some complex of individual beliefs and relations between individuals; group rationality is the result of individual behavior that constitutes social structure. Yet our *explanations* may well invoke group beliefs, rationality, and the like in an irreducible manner. As we have seen may times, higher-level explanations may be multiply realized in lower-level events that are context sensitive. Arguing for group rationality thus does not commit us to group minds just as irreducible biological signals discussed in chapter 4 do not commit us to vital forces.

If such ontological worries are groundless, it might nonetheless seem that an important, related epistemic requirement *is* violated. Put bluntly, the claim is that rationality, justification, and knowledge require that the relevant beliefs and evidence be simultaneously present to a single consciousness. This demand is perhaps most famously found in Descartes's assertion that ideas are knowledge only if they are "clear and distinct to the attending mind." And, of course, philosophers since Plato have assumed that knowledge requires belief and justification in the same individual. So perhaps the problem with *sui generis* collective rationality is not ontological but epistemic.

A first reply to this worry is that it presupposes an internalist account of epistemic notions: it assumes that justification requires that the justification be accessible to the subject. But reliabilist accounts of justification deny precisely this claim. I don't want to get bogged down in this dispute, which seems to depend on overly strong assumptions about conceptual analysis in the first place. My point is only that reliabilist accounts are perfectly respectable and that they do not require the relevant beliefs and evidence be present to consciousness. Instead, they insist only that practice in question reliably produce true beliefs. However, collective practices can do precisely that.

Moreover, the demand that the relevant evidence be present to consciousness runs into a much deeper problem: individual human subjects have a notoriously difficult time meeting it as well. Descartes was bothered by long chains of reasoning, precisely because it is not clear that an individual can have all the steps in a complex proof present simultaneously. Of course, much the same holds for many scientific beliefs, for example—the relevant evidence far outstrips any one individual's ability to recall it or even learn and understand it. We can, of course, read the requirement as only that the beliefs and evidence must be there *implicitly*. Yet making this concession threatens to make collective rationality respectable in principle as well, for the needed beliefs and evidence are also implicit in the community as a whole. The large laboratory that knows some result may have no one individual who can consciously consider and evaluate the total evidence; yet the group as a whole does implicitly have that evidence, for it can consult the relevant specialist within the group. So either internalist demands for accessibility are implausible or group rationality can meet them.

Individualists can, nonetheless, try other strategies to preserve the demand that knowledge requires accessibility to a single consciousness. The traditional epistemic project since Descartes has tried to show that memory and other processes can be sufficient stand-ins for actual presence to the attending mind. For example, if I know that memory is a rational process, then I need not recall the entire proof for some result, only remember that I saw that the previous steps were justified. However, as we shall see at the end of this section, there is an enormous obstacle to any such project, namely, the role authority and trust play for *real* epistemic agents.

So these conceptual objections to collective rationality are unconvincing. I would suggest that the interesting and serious issues about collective rationality are really empirical: they concern whether the potential obstacles to reducing the collective aspects of rationality are empirically real. I turn now to argue that they may be.

Foundationalism's demise opens the way. If rationality does not consist in following universal, a priori rules and if epistemology itself is an empirical discipline, then evaluating belief systems will be done in naturalistic terms. We will want to know about the causes of our beliefs and what that tells us about their veracity. In short, the failure of foundationalism makes possible reliabilist-style, means-ends assessments of rationality. We can ask whether particular practices raise the odds of finding the truth; practices are rational in one sense, then, if they get us the truth, get it better than available alternatives, and so on.

How does this naturalistic epistemology support our claim that rationality can apply to collectives? Clearly it should make us suspicious of arguments that there is some *conceptual* reason why rationality cannot really apply to collectives. How we should understand rationality is an empirical matter, and thus so is the question of what can be rational.

More important, this naturalized view makes the multiple realizations problem likely. On this view, a rational collective is one whose practices promote attainment of the truth. However, there may be many different ways for a group to obtain the truth or other epistemic ends. To use a real-life example, consider the navigation of a ship (see Hutchins 1995). Knowing and following the correct heading is accomplished by distributing numerous specialized tasks to individuals on the ship. A rational cognitive organization is one that produces good outcomes, but (1) that can be done in different ways by differently dividing up tasks and (2) individuals performing those tasks can be entirely nonrational in that they act out of habit or have no idea how the system fits together. All sorts of individual behavior and motives are compatible with a rational community if the community structure brings about epistemic ends. So a rational community need not be a community of rational individuals.

I should point out that it is not just analytic epistemologists who assume that a rational group must be a group of rational individuals. That assumption likewise pervades work from very different perspectives, ones allegedly more cognizant of the social. For example, consider the social constructivist position as advocated by Bruno Latour (1987) in *Science in Action*. Latour outlines in interesting detail the process by which individual scientists "negotiate" with others to build "networks"—how they achieve the acceptance and influence of their own results. Though Latour's is allegedly a social approach to knowledge, its explanatory strategy is individualist through and through. Social structure, groups, institutions, and hierarchies of power make no appearance. Equally important, Latour assumes that if individuals are driven by nonrational factors like influence, then the community to which they belong must be nonrational as well. Latour misses entirely the possibility pointed to above: that nonrational processes at the individual level may produce rational outcomes at the group level. The individualism is obvious. Of course, no such happy outcome is guaranteed, and it takes a close look at actual social processes to determine if they are in fact good ways at getting at the truth. Yet Latour provides no such analysis, and it is therein that his individualism lies. So, here as before, the moral is that rationality at the level of the group need not relate in any automatic fashion to individual rationality.

Let's turn now to the second potential problem for reduction, namely, that collective rationality need not supervene on individual rationality in any straightforward way. In other words, a group of rational individuals is not automatically a rational group.[3] Let me flesh out this possibility with a real example: Imagine two communities, both composed of rational individuals in the sense that all individuals follow standard norms of statistical inference—they reject hypotheses below a certain p value and continue to test unrejected hypotheses. In one group, community opinion is determined, as it were, by majority vote; all studies are published, and then community opinion—as reflected in review articles or the like—depends on whether the hypothesis has been rejected more often than not. In the second group, community opinion is determined by metanalysis: data are pooled and evaluated as a group, then reflected in a review article. As is easy to show, the group that operates by majority opinion may well err more often than that which depends on metanalysis. For example, studies with low power may produce a majority vote against the hypothesis when it is in fact true. Aggregate data analysis can help avoid the error. So the crucial factor is community structure—how *community* opinion is formed.

This example, I would argue, is typical: individual rationality promotes group rationality only in the right context, a context that is in large determined by social institutions and structures. A rational community, as I am using the term here, is one that follows reliable methods. Reliable methods are those that best promote epistemic ends, particularly truth, given the constraints faced by the community. If we take a rational individual to be one who likewise uses reliable methods, then it is clear that a group of rational individuals will produce a rational group only in specific circumstances. More specifically, the social organization of the community must be right. Whether individual agents pursuing the truth provide the best way for the community to get the truth depends on such things as

1. How resources are distributed: When there are economies of scale, indivisibilities, start-up costs, different possible strategies with different costs and payoffs, and so on, then outcomes will vary with how such factors are distributed, even if every individual does the best with what he or she has.[4]
2. The costs and distribution of information: the best community outcomes may require individuals to consider the strategies of others, their likelihood of success, their costs as well as the reliability of results produced by others, the cost of reproducing results oneself, and the likelihood of success of one's own strategies. In the real world, information is not costless nor homogeneously distributed. Social processes—group membership, group status and power, and the like—arguably influence these factors, and some will produce better outcomes than others. Once again a group of rational individuals may not produce the best achievable collective outcome.

I would argue that these problems are ones that confront most "invisible hand" arguments. Good collective results come from the pursuit of individual goals only given some appropriate background social structure. I would also suggest that these conclusions are generalizable to other cases—for example, to collective outcomes produced by individuals who balance truth against credit and other such "sullied" motives or even to collective outcomes produced by individuals trying to do what is best for the community knowledge, not their own garnering of the truth.

So far I have argued that collective rationality is a coherent notion, that rational group practices may be multiply realized in individual behavior, and that individual rationality suffices for group rationality only given the right institutional background. Thus, there is a prima facie case that group rationality is irreducible. I turn now to strengthen that conclusion by arguing that explanations of individual rationality themselves cannot be fully done in purely individualist terms. In other words, I will argue that not just the collective consequences of individual rationality must be explained socially but that individual rationality must be so as well.

If we take the reliabilist line that rationality is a matter of what promotes epistemic goals, then the social will be essential. Social factors influence individual belief. Assuming that I have shown in earlier chapters that such factors sometimes cannot be fully explained in individualist terms, then we must evaluate social influences in determining individual reliability. The key question is then whether particular social factors promote true beliefs or other epistemic goals. In short, reliabilism plus the failure of methodological individualism makes the social essential. Assessing individual rationality requires social information.

Our criticisms above of rationality as a purely formal notion give us another reason to think reliability will be social in nature. To evaluate rationality we must build in substantive information, including in some cases *social* information. Here is a clear example: Someone writing a review article realizes, as we saw above, that the one study, one vote procedure is misguided on statistical grounds. In other words, there are good formal statistical reasons for thinking that such a procedure is flawed. So they instead do a metanalysis, a superior approach on purely formal grounds. Yet the metanalysis itself is only as good as the substantive assumptions underlying it, and those include sociological considerations about science itself. In particular, they depend on knowledge about the publishing practices of journals and the incentive structure for researchers. How often do studies go unpublished because of limits to journal space, editorial politics, or norms in universities and editorial boards about the insignificance of negative results? Answers to these questions are essential for a reliable metanalysis. In short, we need a sociology of scientific practice to do a decent metanalysis.

Some such research has been done. For example, Mirowski (1992, 1994, 1995) has discussed similar social processes involved in determining the fundamental constants in physics, and some metanalysts are aware of the issue (Henrion and Fishoff 1986). Yet such factors are still largely ignored.

So even when scientific reasoning most closely approximates the formalist ideal, judgments about reliability may depend on social information. However, when scientific reasoning is obviously much less formal and mechanical—as usually is the case—then social information has an even more obvious role. In many situations individuals are forced to rely extensively on others for essential information; as others (e.g., Hardwig) have pointed out, much science is sufficiently specialized and collaborative that no one individual can be a reliable producer of truth alone nor can one be in control of all the information needed to make reliable judgments.

The dependence here comes in several different forms, and with different import. First, an individual who develops beliefs based on the information of others may be reliable only if he or she takes social factors into consideration. Why? Relying on the expertise of others without grasping the sociology of the profession may lead to unreliable results—by, most simply, being unable to distinguish the quacks from the real experts. Second and more fundamental, reliance upon experts may be inescapable. If I use social factors to judge experts, it might still be that those judgments are entirely based on my own, individual observation. However, if reliance on social authority is unavoidable—if I make ineliminable use of authority—then individual rationality is strongly social. However, we have good evidence—from the history and sociology of science to everyday experience—that shows just how extensive and thoroughgoing our epistemic dependence is.

How far this dependence goes is thus, I would argue, an empirical issue. Nonetheless, many—both defenders and critics—think the issue is conceptual; they argue that either individual dependence is an a priori fact about the human situation or that it is a conceptual truth that knowledge precludes such dependence. I finish my argument by rejecting both claims.

Some (Coady 1992) argue, for example, that if all reliance on authority were reducible to individual knowledge, then all testimony could possibly be false. But, the argument continues, that is impossible for any knowledge expressible in a language. Others (Foley 1994) argue that I must have intellectual faith in my own cognitive abilities, and that since others are in principle no different from me, I must have such faith in others. Individualists in turn argue that reliance on authority must ultimately be individually "calibrated" or otherwise verified (see Kitcher 1993). Failure to do so, they claim, means epistemic failure.

All these arguments should be resisted. Other people are different from me in some ways, ways that sometimes suggest they are entirely unreliable. Certainly testimony can be entirely false in restricted domains. Perhaps it can be proven a priori that universal falsehood is impossible, though I doubt it. Yet even if such a proof were forthcoming, it would only show that the social must have *some sort* of involvement in individual rationality. It would not tell us how deep that commitment went, where the social was essential and where it was not, and so on. The interesting questions would remain.

Much the same holds for the idea that experts and social authority must be calibrated—justified in individualist terms—for individuals to be rational. If in some domain reliance on authority is ubiquitous and that domain is apparently a paradigm of knowledge and rationality, then we have two choices: reject the philosopher's a priori requirement or become skeptics. But surely the philosopher's a priori stricture has less prior plausibility than skepticism. After all, we have good philosophical grounds for resisting conceptual a priori truths—running from Quine (1961) to Stich (1990)—and an enormous batch of ordinary evidence for rejecting skepticism. Of course the skepticism issue will not really go away this easily. But showing that conceptual, individualist accounts are easily forced to skepticism surely counts against them.

Thus, I conclude that the issues surrounding epistemic individualism are really empirical ones. They depend on results from sociology of knowledge, cognitive science, naturalized philosophy of science, methodology studies, and the like. I think we know enough now about the practice of science to argue that in some areas reliance on authority and trustworthiness of social factors is practically unavoidable. But how deep that reliance goes and how important a role it plays in the total epistemic picture remains an open question.

No doubt I have not proven that epistemic individualism is inevitably misguided—the issues are far too complex to be settled by such a short discussion. Still, the above considerations should both defuse conceptual objections and provide sufficient suggestive evidence to make epistemic holism a live option.

Individualism and Rational Choice

Rational-choice explanations in neoclassical economics have much in common with traditional individualist accounts of epistemic rationality. Both begin from some very general a priori truths or axioms about individual rationality. Both presume that everything important about collective phenomena can be deduced from these universal truths about individuals. I argue in this section that the similarities do not stop there. Neoclassical rational-choice explanations face the same general problems that confront epistemic individualism: collective phenomena cannot be fully captured and, more damaging, individual choice cannot be fully explained in individualist terms. Below I sketch some general arguments for these claims and then use work on human capital theory, rational expectations, and game theory to show the obstacles to individualism in practice. My conclusion is *not* that rational-choice theory has nothing to offer. Appropriately embedded in an overarching social theory, rational-choice explanations can be enlightening indeed. However, the neoclassical tradition suppresses that social structure, leaving it fundamentally flawed.

Chapters 2 and 3 pointed out that social processes, structures, and institutions are frequently multiply realized: diverse individual behavior can bring them about. That fact, we saw, blocks individualism as a reductionist thesis, leaving social explanations a fundamental place. However, multiple realizations are a real prospect in economics where rational choice is supposedly at its strongest. Showing this is, of course, an empirical project. Yet there are some apparent facts about economic processes that make multiple realizations antecedently likely. In particular the fact that economic entities compete, with some winning and others losing, makes multiple realizations quite likely. Competition is often between *organizations.* Firms are founded, transformed, and eliminated in competition with other firms, and it is just such competition that is often invoked to explain why firms must maximize profits. Yet this economic natural selection, like natural selection itself, works on outcomes. If two firms achieve the same result in different ways, economic selection will not "care." Given creativity on the individual level as well as differences in history and culture, we should not be surprised to find that firms solve similar problems in different ways. Profit-maximizing firms may thus be brought about by divergent kinds of individual behavior.

This prospect is more than idle sociological speculation. Recent results in neoclassical tradition support it as well. I have in mind the recent work on the firm discussed in the previous chapter. That work and much like it shows that the traits of firms can result from quite different processes on the individual level. We saw that profit-maximizing, hierarchical firms with long-term employment, internal labor markets, and compensation tied to authority and seniority have many different plausible individualist explanations. We also saw that minimizing transaction costs, maximizing managerial productivity, and maximizing one's value in the managerial market can in theory bring these structures about. Many other explanations are also available.

The upshot is thus that various plausible processes at the individual level are compatible with the traits of firms. But since economic selection operates largely on firms, this means that we should not be surprised to see firm structure realized in diverse individual behavior. In short, even if rational-choice explanations are entirely successful in their own terms, they will still not tell us everything we want to know. Causal patterns at the social level will go unnoticed.

Aside from this multiple realizations problem, neoclassical rational-choice theory faces a variety of other problems if it claims to fully explain in individualist terms. We saw in the last section that a priori constraints do not get us very far in assessing rationality; substantive, often social information is crucial. This same problem, I will argue, confronts rational choice explanation. Determinate results come from rational-choice theory only by implicitly presupposing other information, frequently information about institutional structure, all the while claiming to be explaining in completely individualist terms. Thus, rational-choice theory illegitimately suppresses social information in explaining even individual behavior. The result is that rational-choice explanations are incomplete, even at the individual level; their partial nature undercuts much of the alleged empirical evidence in their favor. In short, rational-choice theory cannot fully explain *individual* behavior.

This abstract argument, if successful, would show rational choice fundamentally misguided qua individualist approach. In the rest of the chapter I flesh out the above argument by looking at three specific pieces of economic theory and research. I begin with human capital theory.

Human capital theory attempts to explain differences in wages and salaries as the result of differential investment in education. Investment in education makes for more productive individuals. Decisions to obtain more education are thus seen as decisions to invest in oneself—hence the name "human capital." Such decisions are analyzed as a standard rational-choice problem under constraints. A crucial factor in such decisions is the return to education. Basically, further education will be pursued if the extra income generated from the education relative to its costs exceeds the interest rate. Differences in wages and salaries are then determined primarily by differential investments in human capital.

A large body of econometric research has tried to estimate the relative return to education, largely by regressing pay differences on differences in education (see Willis 1986).

The above story seems to be a paradigm individualist account, for the large-scale distribution of income is explained on the basis of individual choice and characteristics (see Blaug 1992, 209). Yet, this apparent individualist success is illusory. Human capital theory is not fully individualist, for social structure is presupposed and left unexplained. Moreover, what human capital theory does explain—return to individual education *given* background social structure—is of dubious validity. Since social structure is all but ignored, empirical tests of human capital theory leave many important variables uncontrolled. That at the very least makes them empirically suspect.

Human capital theory rests on at least the following four social presuppositions or givens:

1. *Preferences of workers*: It is assumed that workers are out to maximize monetary gain. In the terminology of the last chapter, a "thick" notion of rationality is presupposed. That means two things: (a) preferences are unexplained, leaving a space for background social factors that explain preference formation and (b) nonmonetary preferences (e.g., for prestige, job satisfaction, authority, and the like) are assumed away, in effect eliminating social factors by fiat. Of course, real-world individuals are influenced by such factors, and every test of human capital that ignores them does so at its peril.
2. *Institutional structure linking pay and productivity*: Human capital theory assumes that increases in education mean increases in productivity, and that differences in productivity determine differences in pay. That picture needs much institutional detail to be persuasive. As we saw in chapter 6, accurately perceiving productivity takes resources, and individuals have incentives to hide their true output. Moreover, if we allow that managers have nonpecuniary motives (authority, prestige, etc.), then they may have reasons to reward employees for things other than output. Many other such facts about the social structure of firms has to be put aside in order to have a straightforward productivity–pay link. Once again, those social factors are potential confounding variables, and ignoring them raises doubts about the positive evidence for human capital theory.
3. *Initial distribution of resources*: Individuals with differential ac-

cess to resources will invest differently in education. So the differential distribution of family wealth, government largesse, and other factors influencing access to education come into play. Again, social structure is presupposed and potentially confounding.

4. *Distribution of jobs*: Here the social assumptions are potentially very powerful and yet almost entirely ignored. Imagine the following dog bone economy (see Sattinger 1993): A truck delivers bones each day to the dog kennel. The bones are dumped in a pile and the dogs fight over them until every dog has the best bone it can get. An empirical study in the mode of the human capital tradition will find a strong correlation between individual dog traits and the size of the bone they get. Yet it is surely an error to think the dog traits are the fundamental or primary determinants of dog bone distribution: the size distribution of bones on the truck is crucially important. Dog traits explain why one dog got this bone rather than that, *given the initial distribution of bones*. It does not explain the latter.

The moral for human capital theory should be obvious. Human capital theory is a supply-side account, taking the distribution of jobs as given. Though that supply might have a fully individualist explanation, it is potentially fertile ground for social explanations as well. Discrimination, class structure, or segmented markets are possible explanations of job distribution. Thus, ignoring these potential institutional facts both leaves human capital theory both incomplete and, once again, empirically challenged. Changes in job distribution clearly are a potentially confounding factor.

I am far from the first individual to criticize human capital theory. At its best, it explains only 30 percent of the variance (see Willis 1986), clearly suggesting other factors are involved. Moreover, most advocates would not deny the role of demand factors, at least if pressed. Yet the actual practice of rational-choice theory remains mired in such individualist excesses. Where the theory is criticized, it is because it ignores the role of differences in individual ability—a criticism that is entirely within the individualist presuppositions described above. Human capital theory continues to be advanced as if it describes the primary factors determining income distribution and were well confirmed. Recent surveys largely ignore the above problems, claiming that it has been "repeatedly confirmed" (Willis 1986, 598) that "wage differentials are based upon differences in productivity which are the result of differences in human capital" (Smith 1994, 75).

Nothing in the above discussion shows that human capital factors are not part of the explanation for wage differences. Rather, the moral is that human capital explanations depend on substantive information about social structure, and thus are explanatorily incomplete and, to the extent that they ignore social variables, poorly tested. Rational-choice theory is not a fully individualist theory and strives to be so only at its peril.

I want to turn now to argue for this general conclusion by looking at two more applications of rational-choice theory: in game theory and in rational expectations in macroeconomics. Both ultimately run into the problems described above. They try to squeeze more from formal rationality than the notion will provide and thus smuggle social structure in the back door to solve the problem, all the while claiming to proceed in individualist terms.

Game theory is currently individualism's best hope in the social sciences. An enormous body of creative work has gone into showing how social phenomena can be explained as the outcome of individual rational choice via the analysis of various games. This work is exciting precisely because it shows how to generalize the rational choice approach outside economics and because it promises to describe processes and mechanisms. Yet game theory falters just when it tries to deduce outcomes from individual rationality.

In one sense, game theory transparently presupposes rather than explains social structure. Game theory analysis begins only after the rules of the game have been set, the preferences of individuals identified, an initial distribution of resources described, and payoffs determined. These are precisely where social institutions and processes have their place.

This objection to game theory may seem relatively minor. After all, many scientific explanations focus on just one part of a complex causal field, and arguably that is all that I have shown about game theory. However, the problem really goes much deeper for two reasons. First, individualists *claim* that game theory shows much more—that game theory fully explains in individualist terms. Second, the presuppositions about the rules of the game and initial distribution are only part of the story. Many of game theory's most impressive and determinate results come only by implicitly assuming much more than the formal notion of rationality delivers. Social factors are once again smuggled in through the back door.

One problem is this: Games sometimes have intuitively many different possible outcomes, all compatible with rational behavior. Many of game theory's most important results come from showing how these "multiple equilibria" can be narrowed down—in short, from demonstrating that one of the many possible solutions is the only one rational agents would reach. It is these determinate results that seem to make game theory so powerful. Yet I would argue that these determinate results cannot be gotten without relying on substantive, controversial assumptions about rationality. In other words, game theorists run into the problem that individualists in epistemology face. Purely formal, a priori universal rules leave us with an impoverished notion of rationality, one insufficient to capture what we want to explain.

This problem is fairly obvious for games that have *focal points* for solutions. There are many games that have multiple Nash equilibria—different sets of strategies that are the best reply to the actions of others. Yet in real life, equilibrium strategies are reached despite their being multiple rational ways to play the game; moreover, some of those solutions are hit upon much more often than others. The cities game is a simple example. Two groups of students are told to divide up a list of cities that includes Washington, Boston, Detroit, Chicago, San Francisco, Seattle, and Los Angeles. Group A is told that its list must include Washington; group B, that its list must include Los Angeles. If the two groups exactly partition the cities (each city appears on one and only one list), they will be given one thousand dollars if they do not, they will get nothing. There are many different possible Nash outcomes here, yet one focal point naturally emerges—that which divides the cities geographically. This outcome is expected because it builds on previous social knowledge or conventions. So analyzing rationality—the best reply—alone gives no determinate result; only bringing in relevant social information provides a solution and explanation. Yet a great many real-life situations resemble the cities game; oligopolic competition, for example, often has many Nash outcomes, yet prior social conventions and knowledge apparently enable corporations to nonetheless pick a single solution. Thus, this problem is thoroughgoing.

Similar, though less obvious problems show up for other attempts to narrow down the possible rational solutions to games. Let me mention three: the assumption that what is rational is "transparent" or that beliefs will be "consistently aligned"; the use of backward induction; and solutions that rest on "trembling hand equilibriums."

Game theorists argue that some outcomes would be chosen over others by bringing in a further assumption about rationality, namely, that reason is transparent. This means that individuals given the same information will infer the same conclusion and, thus, whatever expectations would be rational for me will equally be rational for you facing the same situation. For example, consider the figure 7.1.[5]

	A1	**A2**	**A3**
B1	100, 100	0, 0	50, 101
B2	50, 0	1, 1	60, 0
B3	0, 300	0, 0	200, 200

Fig. 7.1 Example from noncooperative game theory illustrating the transparency assumption.

Imagine B reasons as follows:

> A expects me to play B3 because she thinks I believe she will play A3. But I plan to play B1, because it is the best reply to A1, the move she will make given her expectations.

Game theorists argue that this outcome would not occur with rational players, for rational players would make the same inferences about the game, given that they had the same knowledge. Thus B would reason as follows:

> I believe A will play A2, because she reasons that I will play B2 on the grounds that I expect him to play A2.

This reasoning is perfectly symmetrical, and we thus get B2, A2 as the outcome.

The transparency assumption is crucial to this argument and many others like it in game theory. Yet if the points of chapter 6 and of the previous section of this chapter are correct, then such solutions rest on an untenable notion of rationality—they rest on the idea that purely formal constraints on rationality ensure the same inferences from the same data. Yet the demise of foundationalism was precisely the death of such ideas. Rationality cannot be captured by formal rules acting mechanically on data or observations. Substantive knowledge, often social in nature, is required to be rational and to assess the rationality of others. In short, the transparency of reason requirement is implausible; game theoretical successes that depend upon it do little to enhance the individualist program.

Backward induction, another common route to determinate results, illustrates similar problems. Backward induction is the process of determining what would be rational for players at the last stage in a dynamic game, then determining what would be rational at stage *n*–1, given the prior analysis of *n*. The problem with such reasoning (see Pettit and Sugden 1989) is that even if all players know they are rational, in the sense of maximizing self-interest, at the first and last stages of a game, it does not follow that they are rational at other stages. Or, to put the point differently, what is rational at one stage need not be rational at others, if people learn from experience. Once again the assumption that rationality is a purely formal matter—that it is fully defined independent of time and place—is at work.

We can illustrate this problem with the iterated prisoner's dilemma. Backward induction argues that in the last game *n*, the rational thing to do is confess for familiar reasons. Yet from that it does not follow that confessing is rational at all previous stages or, in other words, that the players are rational for all plays of the game. If I can convince the other player I am playing tit for tat regardless of its rationality at the end of the game, then we both may persist in tit for tat (and be better off because of it) until the game is nearly at an end and tit for tat no longer is possible. So the rational choice in the iterated prisoner's dilemma is not simply deducible from backward induction. Information about payoffs, the frequency of play, the prevalence of norms of cooperation, and other social factors are relevant. The bare facts about individual rationality will not suffice to explain.

One way to deal with deviations from rationality invokes the notion of "trembling hand equilibriums." To show that multiple possible Nash equilibria can be narrowed down to one determinate result, game theorists have invoked the idea that any good equilibrium should be stable. In particular, it should be stable in the face of random deviations from rationality—nervous errors such as a trembling hand—in the play of the game. So it is common to argue that one equilibrium is the solution to a game on the grounds that it is a trembling hand equilibrium. However, for many games the distribution and frequency of trembles or deviations from rationality are crucial parameters; a unique equilibrium is guaranteed only if the frequency and distribution of trembles take on certain values (see Hargreaves-Heap and Varoufakis 1995). In short, deviations from rationality have to have a specific sort of *systematic character*. Thus determinate results are not derived from rationality alone.

These three examples are only a small part of game theory. Yet I would argue that their implicit assumptions about social structure are common throughout the field. Janssen (1996), for example, argues that both evolutionary game theory and Bayesian learning in dynamic games presuppose prior institutional structure. For the Bayesian approach, that structure shows up in the assumption that agents begin with sufficiently similar common priors. For evolutionary game theory, social structure comes in the form of the organization or social environment that does the selecting of strategies.

The moral from all these examples is not that game theory has nothing to contribute. It is instead that game theory provides little support for the individualist program, because social context plays an essential role in individual rationality. When that social context is acknowledged and incorporated, game theory can provide useful insights. Olstrom's (1990) work, for example, on how common-pool resources are really managed is a shining instance: she does not simply conclude that unlimited exploitation is an inevitable fact of reason but instead looks at actual social processes that sometimes allow common-pool resources to be successfully managed. Game theory provides an important part of her story, but only part.

I want to turn now to one final example that illustrates the difficulties of the rational-choice program, namely, the rational-expectations program in economics. Once again the problem is that formal constraints on rationality give us indeterminate results; determinate results come only from illegitimately smuggling in social structure.

The rational-expectations approach has dominated macroeconomics since the early 1980s, and its leading advocate, Lucas, garnered the Nobel prize in 1995. The basic idea behind rational expectations is straightforward: apply the rational maximizing paradigm to *all* individual decisions, including deciding what to believe about the future course of economic variables. Thus, individuals act rationally not only when they respond to current prices but also when they form expectations of future ones. Defenders of this view even go so far as to argue that no macroeconomic model is adequate until it has supplied microfoundations in terms of rational expectations. My target here is not that claim, which was discussed and criticized in chapter 2. Rather, I aim to show that rational-expectations models cannot adequately explain even individual behavior, at least not in individualist terms.

The idea that individuals have rational expectations is given several different glosses. They include the following claims:

1. Individuals learn from errors.
2. Individuals use all available information.
3. Individuals use information up to the point where the marginal return equals the marginal cost of doing so.
4. Individuals do not make systematic mistakes.
5. Individuals form expectations using (or as if they knew) the true structural model of the economy.
6. Individual subjective expectations are equal to the objective expectations, that is those that are the best estimate conditional on the available information using the true structural model.
7. Individual subjective expectations are true.

These claims can further vary depending on whether they are about all individuals, aggregate expectations, representative individuals, long run and short run, and other such parameters. Though the above claims are often used interchangeably, they are clearly logically independent, as will be clear from the discussion to follow.

Social structure is implicitly brought into rational expectations in several ways. Most rational-expectations models assume that individuals use all available information. As should be obvious, this demands much more than simply that they learn from errors or that they gather information so long as the benefits outweigh the costs. In effect, two crucial requirements are involved: (1) Information is assumed to be costless; if information had costs, then individuals would equate marginal revenues with marginal costs just as they would with any commodity or input. (2) Individuals have identical information; they all have access to "all" the information.

Information is seldom costless and, thus, not evenly distributed. However, appeal to social processes—for example, institutions, norms, command over resources, and social networks—is arguably crucial to explaining how, when, and by whom information is gathered when it is costly. Hence, the standard rational expectations assumption that information is costless removes social structure by fiat.

Social structure appears again when we realize that knowing the true model or the best estimate of future variables involves knowing the expectations of others. Some critics seem to regard the regress of expectations—the fact that to form my expectations, I must know yours and vice versa—as an almost conceptual proof that rational expectations must be incoherent. I think the problem really lies elsewhere, for it is clear that in practice the regress is ended. However, various results (see Frydman 1981, for example) strongly suggest that price signals and a firm or consumers' information about their own cost structure alone will not ensure consensus is reached; individuals need not "learn their way" to an average expectation simply by maximizing on price and cost information. Instead, it is arguably social factors that explain the process. Businesses, consumers, and workers develop norms, practices, cultures, networks, organizations, traditions, and so on that allow them to generalize beyond the information contained in prices to form workable estimates of aggregate expectations. Simply to assume accurate estimates of average expectations is to build in social processes from the start.

Other strong evidence that rational expectations assume away social processes comes from the "no trade" results (see Tirole 1982). These results show that in a rational-expectations world, agents have no incentive to trade. The market as a dynamic social process that conveys information disappears, because that information has been built in from the beginning. The common knowledge and prealigned plans built into rational expectations removes the reasons for trade.

The claim that individual subjective expectations of future variables equals their objective expectation also belies a very strong notion of rationality, one that depends on denying the points of the first section. The first section argued that epistemic rationality is not a purely formal matter. This conclusion entails that the relationship between data and hypothesis is not a purely formal one; purely logical, conceptual rules will not suffice to tell you what hypothesis to believe, given the data. Yet the rational-expectations hypothesis denies this. It assumes that the question "What is the true objective expectation of the hypothesis, given the data?" is something decidable on purely formal grounds, primarily by the rules of statistical inference. I want to argue that this assumption fails; rational expectations are not a matter of formal rationality.

Now in one sense this is entirely obvious. Rational-expectations models typically assume agents know the "true" macroeconomic model (or, better, the model being advocated), at least in its basic structure. It is this specification of variables that is then used to derive the "objective" expectations, given the data. Though advocates of rational expectations are upfront about this procedure, it is quite at odds with the advertised virtues of rational-expectations approach. Proponents claim that it is the natural extension of rational choice under constraints to expectations themselves. The idea that expectations should fit the rational maximizing model so fruitful elsewhere has great rhetorical appeal, especially when the obvious failings of adaptive expectations are pointed out. Yet that appeal is really gotten on the cheap. It is not rational maximizing that is doing the work here, it is the *assumption* that individuals have the true macroeconomic model of the economy. This is much like trying to defend the rationality of the scientific enterprise by first assuming that scientists have the correct model of the world and then proving, low and behold, that they make fairly accurate predictions (according to the model already assumed!).

I think, however, the problems run deeper than the fact that the rational-expectations practice is at odds with its rational-choice preaching or the obviously lurking issue of falsifiability. *Rational expectations are not a matter of formal rationality even assuming agents have the true structural model of the economy.* Even its more careful critics have bought into this last step of the argument. Pesaran (1987), for example, says that "clearly, when rational expectations are formed on the basis of the true model of the economy, the subjective probability distributions held by economic agents will be the same as the objective." ("Objective" here means the expectations that economists would form using the true model and available data, not the literally true objective probabilities in the world.) However, we have good philosophical and practical reasons to reject that inference.

The philosophical reasons concern the place and scope of statistical inference. Our earlier claims about rationality suggest that statistical inference is not a purely formal matter—that substantive empirical assumptions about the world are involved. I would argue that this is still the case even if we have a correct specification of the macroeconomic variables. Estimating the values of those variables is still not a purely formal matter, and thus inferring the rational expectation cannot be done on purely formal, logical grounds. To establish that thesis would of course take a long story about the foundations of statistical inference. But the following three considerations should suffice to make the point plausible:

1. If economic agents with the true model pick estimates that are consistent and unbiased in the traditional statistical sense, they will be using criteria that have "no merit" according to some statisticians (see Howson and Urbach, 1993, 233).

2. If economic agents infer on the basis of least square regression analysis, they will be making irrational inferences according to some statisticians.
3. If economic agents infer without reference to their subjective priors they will be making irrational inferences according to some statisticians.

In short, are our economic agents defenders of Bayesian approaches to the foundations of statistics or some more classical approach? If the latter, which instantiation?

I would argue, following the general claims about rationality earlier, that the classical picture of statistical inference is misguided and that Bayesian approaches better capture the real empirical elements crucial to statistical inference. However, I need not establish that claim, only that a rational economic agent ought to take the Bayesians into consideration. As soon as he or she does that, then the idea that there is one correct estimate of future variables looks dubious indeed. In effect, the rational consumer of statistical theory will be making choices in the face of ignorance, choices that an exhaustive literature shows have no obvious formal rational solution.

More practical considerations also argue for this conclusion. Given two econometricians, both of the classical persuasion about statistics, a data set, and a basic structural model, will they reach the same estimates of the variables? If inference is purely a formal, rational matter, they should. But we have substantial empirical evidence that they do not, both anecdotal and systematic.

Hence, the rational-choice approach again cannot be the full explanation of individual behavior. This conclusion does not *prove* that the best explanation of individual expectations must be at least partly in social terms, but it surely does show that the appeals of purely individualist accounts are vastly overrated.

In the end we can draw the same moral about rational knowers and rational agents: if we give up on the idea that rationality is determined and exhausted by a priori rules, then any adequate account must look at how real agents deal with the world. As the previous chapters have argued at length, we cannot fully understand those agents in purely individual terms. Therefore, it is not surprising to find that even in the domain of rationality—individualism's stronghold—the social has an essential role to play.

8

Reviewing the Argument

My argument against reductionism and its individualist variants has been a long one. In this last chapter I pull the claims of the previous chapters together to summarize the overall argument.

How the sciences relate, I have argued, is ultimately an empirical issue. Broad conceptual facts about the primacy of the physical, about human agency, or even about the ontology of society will not tell us how the sciences must relate. Like other failed attempts to show a priori how human inquiry must go, arguments of this sort give philosophical or conceptual considerations a power far beyond their means.

If we look at real sciences and how they interrelate, we will see that the traditional reductionist picture badly describes the actual practice of science. Some of our best examples of tying higher-level explanations to their lower-level counterparts—for example, molecular biology and thermodynamics—do not involve reductions that show one theory can do all the explanatory work of the other. Instead, we see much more complex interconnections that leave higher-level explanations an essential role.

Reduction fails in these cases for two basic reasons. First, theories at different levels can divide up the world differently. One and the same complex phenomena can be categorized in different ways by theories with different predicates or categories, and causal patterns discernible in one set of terms need not be perspicuously captured in another. We can, for example, say much about the causal processes driving the immune system, but that causal knowledge is through and through biological. Second, reduction also fails because even when we explain in terms of lower-level entities—in terms of molecules for example—we often do so by using higher-level terms such as "antibodies." In short, higher-level accounts often are not just autonomous but essential. I have no conceptual proof that things must always work this way, but I have given plenty of empirical evidence that they often do.

Philosophers are sometimes wary of this "unity in diversity" picture because they believe it has untoward ontological or metaphysical implications. I argue that it does not; moreover, given how clearly the actual practice of science supports my nonreductive view, I am much more inclined to reject abstract metaphysical arguments than I am to accept claims that the irreducible special sciences should be eliminated, that they describe no real causal processes, and the like. Multiple levels of explanation need not commit us to multiple levels of independent entities, for the fact that A's are composed of B's entails nothing about what categories we must use to explain A or B. So the nonreductive unity I defend need not commit us to vital forces or group minds; it is compatible with, but does not require, a thoroughgoing physicalism as an *ontological* doctrine.

I have also argued that the reductionist program cannot be saved by switching to claims about *explanation*. The root notion behind reduction is that one theory can explain everything another can and more. But if lower-level accounts serve to fully explain, then either reduction must be possible after all or any higher-level account must be explanatorily otiose. However, we have good reasons to doubt that the latter is generally the case, as molecular biology and thermodynamics amply illustrate. Both seem to be nonreducible and yet are great scientific successes. Yet, if we grant that strict reduction fails, then lower-level accounts will be explanatorily incomplete for they potentially leave many questions(for example about aggregate causal processes—unanswered.

There are versions of reductionism that claim still less, yet they too are implausible. There is no general demand for lower-level mechanisms if we are to have good science. Some of our best science explains in ignorance of details at lower levels. Nor is information about lower-level processes always crucial. When (1) our understanding of mechanisms is minimal, (2) our knowledge of higher-level processes is strong, or (3) our accounts of higher-level processes presuppose little about lower-level detail, then the demand for mechanisms carries little force. Evidence about mechanisms can surely sometimes be important, but there is no reason it must always be. Nor is there compelling reason to think that proceeding in lower-level terms is the best strategy to get at the truth. Much history of science seems to show quite the opposite, and we have good reasons why that should be so—patterns captured by higher-level descriptions are likely to be missed.

Of course, reductionist intuitions are not completely misguided.

Complex wholes obviously do not exist nor act independently of their parts. Information about underlying detail can help confirm and can deepen our explanations. But these are anemic reductionist doctrines that leave space for autonomous and essential higher-level explanations.

It is this general picture that I illustrate and defend in looking at individualism in the social sciences. Social processes and social institutions do not exist nor act apart from individuals. However, these truisms say little about how to explain social processes. Social explanations—explanations in terms of social entities and processes—may pick out patterns that can be brought about in diverse individual behavior. Many explanations of individual behavior depend upon or presuppose social information about institutions, and the like. In each case reduction is thwarted. Thus we have empirical reasons for doubting that good social explanations can be fully captured in purely individualist terms.

I have also tried to show that logically weaker forms of individualism are no more plausible. As we saw above, individualism as a doctrine about mechanisms has no warrant in general maxims about good science. So perhaps claims that individualist explanation is superior or at least necessary must have their basis in facts special to the social sciences. That possibility leads us to look in detail at the claim that explanations in terms of individuals are preferable because they provide the *best* explanation of social phenomena. Unfortunately, appeals to explanatory virtues in the abstract do not get us far, yet those making such claims seldom are very explicit about what explanation involves. When we turn to look at the specific claims of individualist theories to better explain, we see that what is doing all the work is not some general traits of good scientific explanation but specific, contingent, substantive, and frequently empirically dubious presuppositions about rationality, optimality, and the like.

While these problems confront the individualist program generally, my main target throughout has been neoclassical economics, or at least certain prominent strands of that doctrine. Neoclassical economics is explicitly individualist in outlook and takes its individualism to count strongly in its favor. That support is vastly overrated. Frequently, neoclassical accounts profess individualism but in fact explain by making use of explanations in nonindividualist terms. Rational choice and game theoretical explanations take as given, and thus do not explain, much institutional structure, something I have tried to trace out in a fair amount of detail. Often the individuals of neoclassical theory are no such thing; they are instead odd aggregate entities such as "representative individuals" or they are social entities such as corporations treated as if they had preferences and the like. Of course, this smacks of group minds and is hardly an "individualist" explanation.

Neoclassical economics comes closest to its individualist aspirations when it invokes very strong or "thick" notions of rationality. Stronger notions of rationality—like those invoked in rational expectations—make fewer explicit assumptions about social structure. Yet they do so only by relying on notions of rationality that are highly implausible, both as descriptions of real agents or even as descriptions of an ideal rationality. As we saw in the last chapter, rational-expectations doctrines assume a link between evidence and hypothesis choice that even ardent believers in a pure inductive logic would hesitate to endorse. Moreover, such thick notions of rationality protect individualism sometimes only at a very high cost—for example, the cost of implying that economic agents never trade.

Though I have only discussed reductionist programs directly in the social sciences and cell biology, the general perspective developed here ought to have implications for a variety of other areas where reductionism comes up. I have already pointed to some of these implications for our understanding of epistemic rationality and the practice of science. The doubts raised about rational-choice theory taken as an individualist program ought also to raise doubts about the individualism implicit in contemporary contractarian approaches to morality and political philosophy. Current discussions of reductionism in the philosophy of mind too often occur without a complete understanding of the obstacles reduction faces elsewhere; I suspect, for example, that many so-called neurological explanations are nothing of the sort in that they implicitly or explicitly presuppose higher-level *cognitive* information. Disputes in ecology sometimes raise issues of holism and reduction, and I suspect that the perspective developed here would help shed light on those issues also. Indeed, a stronger argument for my main theses would have looked at these areas in detail as well. That, however, is a project for another time—perhaps another lifetime.

Notes

Chapter 2

1. For a discussion of some real possibilities along these lines as well as the best work in economics on how a realistic model of economic selection might be built, see Nelson and Winter (1982).
2. For an overview of both theory and empirical work on individual consumption behavior with frequent reference to macroeconomic variables, see Deaton and Muellbauer (1980).
3. For a discussion of some of the complexities, see Nelson (1989).
4. For a survey of the issues, see Weintraub (1979).
5. For a survey that motivates rational expectations in part by the demand for mechanisms, see Begg (1982).
6. For example, by Laudan (1987).

Chapter 3

1. Note that such indeterminacy does not contradict my earlier commitment to the supervenience of the social on the individual. Supervenience, as I formulated it, claims that once all the facts about individuals are set, then so too are the social facts. But this does not entail that each individual fact or kind of fact uniquely determines some social fact or kind (except for the trivial case where one counts all the individual facts as one disjunctive fact or kind). The problem I am raising in effect argues that determination or supervenience relations between the social and individual must be relatively global, as opposed to local, in nature if they are to be plausible.
2. It might seem that specifying the beliefs involved would suffice. However, I doubt that this is a general solution for at least three reasons: (1) beliefs, desires, and mental states generally connect with the same act, physically described, in indefinitely many ways; (2) the social significance of individual be-

havior often goes beyond and does not depend on the intentions of the actor, thus giving acts different social specifications even when belief is specified and invariant; and (3) it may well be that individuating and/or identifying beliefs cannot be done without reference to the social context.

3. Note that the kind of case Mellor cites must be distinguished from the situation where the reducing theory corrects the reduced theory. In the latter case, the laws of the higher-level theory are known or shown to be partially inaccurate while the reducing theories avoid such inaccuracies. While we do have reduction here by means of only approximate connections between predicates, the approximation occurs because the social terms and laws are in fact misapplied and inaccurate in their own domain—so, reducing theory does everything the reduced theory does and does it better. However, when higher-level laws and predicates are known to accurately apply to their domain and are known to be only approximately correlated with lower-level predicates, we have an entirely different situation—the reducing theory cannot handle all events at the higher level.

4. In actual practice, one theory comes to do the explanatory work of another in much more complicated ways: some higher-level kinds are dropped altogether (because the laws invoking them are false), others are corrected and then equated with lower-level predicates, and others may be directly reduced without alteration. Although my argument simplifies greatly, the simplification makes no difference so long as there are some explanations involving social kinds that are successful.

5. Either (2) can be taken as the number of questions if we count every different specification of contrast classes and answer kinds as determining a new question, or (2) can be phrased in terms of the number of questions answered and the range of contrast classes and answer kinds handled for each question. Popper (1972) has also suggested the breadth of questions answered as a dimension for theory assessment.

6. Hull and Richardson try to support their weakened version of reduction by arguing that molecular genetics has clearly reduced classical genetics but has done so without giving the bridge-laws demanded by traditional accounts of reduction. However, as Patricia Kitcher has pointed out (1982), it is doubtful that molecular genetics really warrants a revision in the traditional account for reasons similar to those considered here: (a) classical genetics on the whole did not provide accurate laws, and thus molecular genetics could explain without providing a derivation of higher-level kinds and laws; and (b) where classical genetics does hold true, molecular biology does not generally replace such laws but rather provides an account of why the law holds true in each particular case. In short, molecular genetics appears to reduce only because there was either no explanatory theory to reduce or because reduction is confused with explanation case by case.

7. Aside from the bare fact that purely social theories do sometimes appear to explain, this thesis also faces more general problems. Unless we can say why

explanation of social wholes differs essentially from macroexplanations in other domains, we seemed forced to the conclusion that no purely macroexplanation, no matter the domain, explains. That would make most of our natural science inadequate as well.

Chapter 4

1. Often the theory to be reduced (the "reduced" theory) is a "higher-level" theory in that it describes entities that are themselves composed of entities described by the reducing theory—the latter is thus a "lower-level" theory. I will thus use higher-level/reduced and lower-level/reducing interchangeably.

2. The problem presented by functional terms is independent of other traditional objections raised by antireductionists. Functional terms describe causal roles in larger systems; they do not necessarily presuppose either (a) that there is some natural selection process producing that function (though there may be) or (b) that teleological systems cannot be analyzed in purely causal terms (the elimination of teleology and the eliminability of functional terms are separate issues). Thus this argument differs from those of Rosenberg (1985) based on (a) or of Jacobs (1986), which appeal to (b).

3. Debates over reducibility typically focus on whether molecular genetics suffices to reduce classical genetics. I avoid that focus here and instead look at the more general question of the relation between biology at the cellular level and biochemistry. Since cell biology involves much more than the mechanics of gene expression and since the status of the classical notion of a "gene" is itself uncertain (see Kitcher 1982), it is useful to put the question in more general terms.

4. Schaffner (1993) argues that the multiple realizability problem can be handled if we understand that basic predicates are not multiply realized; diversity comes in the range of initial conditions for applying lower-level or biochemical predicates. This just pushes the problem off to another area. If the range of initial conditions is completely open ended and thus not capturable in biochemical terms, then the biological has an essential role to play in picking out the real generalization. I could define "chair" physically by identifying it as "a physical thing," and then argue that it is only a matter of specifying the initial conditions in physical terms that remains to be done. Such a solution to the multiple-realizations problem obviously is no reduction. Of course, even if the appeal to initial conditions solved the mutiple-realizations problem, it would leave untouched the other difficulties of context sensitivity and presupposing higher-level information.

5. Robinson (1992) is critical of my doubts about biochemistry, particularly my heuristic claims. Yet so far as I can see, Robinson really has no deep disagreement: he grants my main claim that biologists are not providing reductions in the traditional sense, and he also acknowledges that biological information

plays a crucial role in looking for biochemical mechanisms. My attack on reductionist approaches is an attack on purely biochemical approaches; I certainly grant Robinson's claim that biochemistry informed by biology has done much for the unity of science.

6. Note that even this explanation is not really purely biochemical; it would be so only if we could give some molecular definition of cellular location, in this case of rough endoplasmic reticulum.

7. This is one reason of several why I do not accept Rosenberg's (1994) claim that biological science must be treated instrumentally—that only science at the level of macromolecules provides real science, that is, explanatory laws. The macromolecular explanations in biology are so tightly interconnected with the biological that I am skeptical that we can treat them differently. I am also suspicious of Rosenberg's claims that biology is fundamentally different from physics in terms of its reliance on idealizations, simplifications, ad hoc rules of thumb, and so on. Following Cartwright (1984), I think these are also rife in physics, and that much of the "predictive improvement" in physics that Rosenberg focuses on comes from their use.

8. Of course, theories with different vocabularies will be trivially compatible unless further assumptions are involved. Compatibility is a nontrivial requirement once underlying mechanisms, supervenience bases, and other cross-theoretic links are established.

9. The account given here is in part inspired by Maull (1977); Darden and Maull (1977); and Kitcher (1984).

Chapter 5

1. Few accounts of causation require that causes be necessary for their effects, especially if we are talking about kind of effects, not instances. Even Lewis's counterfactual account, which does make causes necessary for their effects (*c* caused *e* iff had *c* not occurred, then *e* would not have) do so only relative to background conditions or causes that specify the possible worlds for evaluating the counterfactual.

Chapter 6

1. See Day and Kincaid (1994) for a much more detailed defense of the claims here made about IBE.

Chapter 7

1. See Solomon (1994) for defense of similar ideas.

2. I assume rather than defend this means-end approach to rationality. I do

not think it exhausts what we can say, but it does follow naturally from the naturalized approach to epistemology that is the successor to foundationalism. Moreover, it is the approach adopted by the rational-choice explanations I aim to attack, so it is a presupposition to which I am entitled.

3. Of course, it might be that not every group of rational individuals was a rational group but that having rational individuals was a necessary condition for group rationality. However, the arguments given above show that more limited condition misguided as well.

4. Here I am turning Kitcher's (1993) interesting models on their head, as it were: he used them primarily to argue that "epistemically sullied" individuals could reach optimal outcomes; I am using some of the same points to show that epistemically pure individuals need not reach optimal community outcomes.

5. Taken from Hargraves-Heap and Varoufakis (1995).

References

Achinstein, Peter. 1983. *The Nature of Explanation*. Oxford: Oxford University Press.

Alberts, B., Bray, D., Lewis, J., Raff, M., Roberts, K., and Watson, J. 1994. *Molecular Biology of the Cell*. New York: Garland Publishing.

Allen, Garland. 1975. *Life Science in the Twentieth Century*. Cambridge: Cambridge University Press.

Arrow, K. 1968. "Mathematical Models in the Social Sciences." In *Readings in the Philosophy of the Social Sciences*, ed. M. Broadbeck. New York: Macmillan.

Ashenfelter, O., and Layard, R., eds. 1986. *Handbook of Labour Economics*. Amsterdam: North Holland.

Bacon, J. 1986. "Supervenience, Necessary Co-extension, and Reducibility." *Philosophical Studies* 49: 163–76.

Batenburg, A., Brasseur, R., Ruysschaert, J., Scharrenburg, G., Slotboom, A., Demel, R., and Krkuijff, B. 1988, "Characterization of the Interfacial Behavior and Structure of the Signal Sequence of *Escherichia coli* Outer Membrane Pore Protein." *Journal of Biological Chemistry* 263: 4202–7.

Bates, F., and Harvey, C. 1975. *The Structure of Social Systems*. New York: Gardner Press.

Becker, Gary. 1976. *The Economic Approach to Human Behavior*. Chicago: University of Chicago Press.

———. 1981. *A Treatise on the Family*. Cambridge: Harvard University Press.

Begg, David. 1982. *The Rational Expectations Revolution in Macroeconomics*. Oxford: Philip Allan.

Belnap, N., and Steel, T. 1976. *The Logic of Questions and Answers*. New Haven: Yale University Press.

Berridge, M. 1985. "The Molecular Basis of Communication Within the Cell." *Scientific American* 253: 142–52.

Bickle, John. 1995. "Psychoneural Reduction of the Genuinely Cognitive: Some Accomplished Facts." *Philosophical Psychology* 8: 265–85.

Blaug, Mark. 1992. *The Methodology of Economics*. Cambridge: Cambridge University Press.

Blobel, G., and Dobberstein, B. 1975. "Transfer of Proteins Across Membranes." *Journal of Cell Biology* 67: 835–62.

Bonjour, Lawrence. 1985. *The Structure of Empirical Knowledge*. Cambridge: Harvard University Press.

Boyd, Richard. 1985. "Lex Orandi est Lex Credendi." In *Images of Science*, ed. Paul Churchland and C. Hooker. Chicago: University of Chicago Press.

Broadbeck, M., ed. 1968. *Readings in the Philosophy of the Social Sciences*. New York: Macmillan.

Burge, T. 1979. "Individualism and the Mental." In *Midwest Studies in Philosophy IV.* Minneapolis: University of Minnesota Press.

Cartwright, Nancy. 1984. *How the Laws of Physics Lie*. Oxford: Oxford University Press.

Chrispeels, M., and Raikhel, N. 1992. "Short Peptide Domains Target Protein to Plant Vacuoles." *Cell* 68: 613–16.

Churchland, Paul, and Hooker, C., eds. 1985. *Images of Science*. Chicago: University of Chicago Press.

Churchland, Paul. 1978. "Eliminative Materialism and Propositional Attitudes." *Journal of Philosophy* 78: 67–91.

Coady, C. A. 1992. *Testimony*. Oxford: Oxford University Press.

Cooley, C. 1956. *Social Organization*. Glencoe, Ill.: Free Press.

Creighton, T. 1984. *Proteins*. New York: W. H. Freeman.

Dalbey, R., and Wickner, W. 1987. "Leader Peptidase of *Escherichia coli*: Critical Role of a Small Domain in Membrane Assembly." *Science* 235: 783–89.

Danto, A. 1973. "Methodological Individualism and Methodological Socialism." In *Modes of Individualism and Collectivism,* ed. J. O'Neill. London: Heineman.

Darden, L., and Maull, N. 1977. "Interfield Theories." *Philosophy of Science* 11: 43–64.

Day, Timothy, and Kincaid, Harold. 1994. "Putting Inference to the Best Explanation in Its Place." *Synthese* 98: 271–95.

Deaton, Angus, and Muellbauer, John. 1980. *Economics and Consumer Behavior*. Cambridge: Cambridge University Press.

Dennett, Daniel. 1969. *Content and Consciousness*. London: Routledge.

Dobb, Maurice. 1973. *Theories of Value and Distribution Since Adam Smith*. Cambridge: Cambridge University Press.

Dore, R. 1973. "Function and Cause." In *The Philosophy of Social Explanation*, ed. Alan Ryan. Oxford: Oxford University Press.

Dray, W. 1964. *Philosophy of History*. Englewood Cliffs, N.J.: Prentice Hall.

Eckstein, Otto. 1983. *The DRI Model of the U.S. Economy*. New York: McGraw Hill.

Elster, Jon. 1985. *Making Sense of Marx*. Cambridge: Cambridge University Press.

———. 1986. "Reply to Comments." *Inquiry* 29: 65–77.
———. 1989. *Nuts and Bolts for the Social Sciences*. Cambridge: Cambridge University Press.
Fair, Ray. 1984. *Specification, Estimation and Analysis of Macroeconomic Models*. Cambridge: Harvard University Press.
Fama, Eugene. 1980. "Agency Problems and the Theory of the Firm." *Journal of Political Economy* 88: 288–307.
Fidelio, G., Austen, B., Chapman, D., and Lucy, J. 1986. "Properties of Signal Sequence Peptides at an Air-Water Interface." *Biochemical Journal* 238: 301–4.
Fjose, A., McGinnis, W., and Gehring, W. 1985. "Isolation of a Homeo Box-Containing Gene from the *Engrailed* Region of Drosophila and the Spatial Distribution of its Transcripts." *Nature* 313: 284–89
Foley, Richard. 1994. "Egoism in Epistemology." In *Socializing Epistemology*, ed. Frederick Schmitt. Lanham, Md.: Rowman & Littlefield.
Friedlander, M., and Blobel, G. 1985. "Bovine Opsin Has More Than One Signal Sequence." *Nature* 318: 338–42.
Friedman, Michael. 1974. "Explanation and Scientific Explanation." *Journal of Philosophy* 71: 5–19.
Frydman, R. 1981. "Sluggish Price Adjustments and the Effectiveness of Monetary Policy Under Rational Expectation." *Journal of Money, Credit, and Banking* 13: 94–102.
Gal, S., and Raikhel, N. 1993. "Protein Sorting in the Endoplasmic System of Plants." *Current Opinion in Cell Biology* 5: 636–40.
Garfinkel, Alan. 1981. *Forms of Explanation*. New Haven: Yale University Press.
Gehring, W. 1985. "The Molecular Basis of Development." *Scientific American* 253: 152–64.
Globus, G., ed. 1976. *Brain and Consciousness*. New York: Plenum.
Goldfarb, D., Gariepy, J., Schoolnik, G., Kornberg, R. 1986. "Synthetic Peptides as Nuclear Localization Signals." *Nature* 322: 641–44.
Goldman, Alvin. 1986. *Epistemology and Cognition*. Cambridge: Harvard University Press.
Gordon, Scott. 1991. *The History and Philosophy of Social Science*. London: Routledge.
Greenway, David, ed. 1989. *Current Issues in Macroeconomics*. London: Macmillan.
Grimes, Thomas. 1995. "The Tweedledum and Tweedledee of Supervenience." In *Supervenience,* ed. Elias Saraellos and Umit Yalcin. Cambridge: Cambridge University Press.
Hannan, M., and Freeman, J. 1989. *Organizatinal Ecology*. Cambridge: Harvard University Press.
Hardwig, J. 1985. "Epistemic Dependence." *Journal of Philosophy* 82: 335–49.

Hargreaves-Heap, Shaun, and Varoufakis, Yanis. 1995. *Game Theory*. London: Routledge.

Harman, Gilbert. 1965. "Inference to the Best Explanation." *Philosophical Review* 74: 88–95.

Haugland, J. 1981. *Mind Design*. Cambridge: MIT Press.

Hay, R., Bohni, P., and Gasser, S. 1984. "How Mitochondria Import Proteins." *Biochimica et Biophysica Acta* 779: 65–87.

Hellman, Geoffrey, and Thompson, F. W. 1975. "Physicalism: Ontology, Determination and Reduction." *Journal of Philosophy* 72: 551–64.

Hempel, Carl. 1965. *Aspects of Scientific Explanation and Other Essays in the Philosophy of Science*. New York: Free Press.

———. 1966. *The Philosophy of the Natural Sciences*. New York: Prentice Hall.

Hill, C. 1984. "In Defense of Type Materialism." *Synthese* 59: 295–321.

Hillier, Brian. 1991. *The Macroeconomic Debate*. Oxford: Basil Blackwell.

Hohfeld, J., and Hartl, F. 1994. "Post-translational Protein Import and Folding." *Current Opinion in Cell Biology* 6: 499–509.

Homans, George. 1974. *Social Behavior: Its Elementary Forms*. New York: Harcourt Brace Jovanovich.

Hoover, Kevin. 1990. *The New Classical Macroeconomics*. New York: Basil Blackwell.

Howson, Colin, and Urbach, Peter. 1993. *Scientific Reasoning: The Bayesian Approach*. LaSalle, Ill.: Open Court.

Hull, D. 1974. *The Philosophy of Biological Science*. Engelwood Cliffs, N.J.: Prentice Hall.

Hutchins, Edwin. 1995. *Cognition in the Wild*. Cambridge, N.J.: MIT Press.

Jacobs, J. 1986. "Teleology and Reduction in Biology." *Biology and Philosophy* 1: 389–401.

Janssen, Maarten. 1996. "Individualism and Equilibrium Coordination in Games." Unpublished ms, Erasmus University, Amsterdam.

Kaiser, C., Preuss, D., Grisafi, P., and Botstein, D. 1987. "Many Random Sequences Functionally Replace the Secretion Signal Sequence of Yeast Invertase." *Science* 235: 312–18.

Kim, Jaegon. 1993. *Supervenience and Mind*. Cambridge: Cambridge University Press.

Kincaid, Harold. 1996. *Philosophical Foundations of the Social Sciences*. Cambridge: Cambridge University Press.

Kitcher, P. 1982. "Genes." *British Journal for the Philosophy of Science*. 33: 337–59

———. 1984. "1953 and All That. A Tale of Two Sciences." *Philosophical Review* 43: 335–74.

———. 1989. "Explanatory Unification and the Causal Structure of the World." In *Scientific Explanation*, ed. Philip Kitcher and Wesley Salmon. Minneapolis: University of Minnesota Press.

———. 1993. *The Advancement of Science: Science Without Legend, Objectivity Without Illusions*. New York: Oxford University Press.

Latour, Bruno. 1987. *Science in Action: How to Follow Scientists and Engineers Through Society*. Cambridge: Harvard University Press.

Laudan, Larry. 1987. "Progress or Rationality? The Prospects for Normative Naturalism." *American Philosophical Quarterly* 24: 19–31.

Lehnhardt, S., Pollit, S., and Inouye, M. 1987. "The Differential Effect on Two Hybrid Proteins of Deletion Mutations Within the Hydrophobic Region of the *Escherichia coli* OmpA Signal Peptid." *Journal of Biological Chemistry* 262: 1716–19.

Lenoir, Timothy. 1982. *Strategies of Life*. Chicago: University of Chicago Press.

Mandelbaum, M. 1973. "Societal Facts." In *Modes of Individualism and Collectivism,* ed. J. O'Neill. London: Heineman.

Martin, M. 1972. "On Explanation in Social Science: Some Recent Work." *Philosophy of Social Science* 2: 67–68.

Maull, N. 1977. "Unifying Science Without Reduction." *Studies in the History and Philosophy of Science* 9: 143–62.

McDonald, Graham, and Pettit, Philip. 1981. *Semantics and Social Science*. London: Routledge and Kegan Paul.

Mellor, D. 1982. "The Reduction of Society." *Philosophy* 57: 51–74.

Nelson, Alan. 1984. "Some Issues Surrounding the Reduction of Macroeconomics to Microeconomics." *Philosophy of Science* 51: 573–94.

———. 1989. "Average Explanation." *Erkenntnis* 30: 23–42.

Nelson, Richard, and Winter, Sydney. 1982. *An Evolutionary Theory of Economic Change*. Cambridge: Harvard University Press.

Ng, D., and Walter, P. 1994. "Protein Translation Across the Endoplasmic Reticulum." *Current Opinion in Cell Biology* 6: 510–16.

Oliver, D. 1985. "Protein Secretion in *Escherichia coli.*" *Annual Review of Microbiology* 39: 615–48.

Olstrom, Eleanor. 1990. *Governing the Commons*. Cambridge: Cambridge University Press.

O'Neill, J., ed. 1973. *Modes of Individualism and Collectivism*. London: Heineman.

Peel, David. 1989. "New Classical Macroeconomics." In *Current Issues in Macroeconomics*, ed. David Greenway. London: Macmillan.

Pesaran, M. H. 1987. *The Limits to Rational Expectations* Oxford: Basil Blackwell.

Pettit, F., and Sugden, R. 1989. "The Paradox of Backward Induction." *Journal of Philosophy* 86:169–82.

Popper, Karl. 1950. *The Open Society and Its Enemies*. Princeton: Princeton University Press.

Powell, P., Kyle, J., Miller, R., Pantano, J., Grubb, J., and Sly, W. 1988. "Rat Liver B-glucuronidase." *Biochemistry Journal* 250: 547–55.

Putnam, Hiliary. 1981."Reductionism and the Nature of Psychology." In *Mind Design*, ed. J. Haugland. Cambridge, Mass.: MIT Press.

Quine, W. V. O. 1961. *From a Logical Point of View*. New York: Harper and Row.

Richardson, Robert. 1979. "Functionalism and Reduction." *Philosophy of Science* 46: 533–58.

Robinson, J. 1986. "Reduction, Explanation, and the Quests of Biological Research." *Philosophy of Science* 53: 333–53.

Robinson, Joseph. 1992. "Discussion: Aims and Achievements of the Reductionist Approach in Biochemistry/Molecular Biology/Cell Biology: A Response to Kincaid." *Philosophy of Science* 59: 465–71.

Rosen, Sherwin. 1982. "Authority, Control and the Distribution of Earnings." *Bell Journal of Economics* 13 (1982): 311–23.

Rosenberg, A. 1985. *The Structure of Biological Science*. Cambridge: Cambridge University Press.

———. 1989. "From Reductionism to Instrumentalism." In *What the Philosophy of Biology Is*, ed. M. Ruse. Dordrecht: Kluwer.

———. 1994. *Instrumental Biology, or The Disunity of Science*. Chicago: University of Chicago Press.

Ruben, David. 1985. *The Metaphysics of the Social World*. London: Routledge and Kegan Paul.

Ruse, M. 1976. "Reduction in Genetics." *PSA* 1974: 653–70.

Ryan, Alan., ed. 1973. *The Philosophy of Social Explanation*. Oxford: Oxford University Press.

Sabatini, D., Kreibaich, G., Morimoto, M., and Adesnik, M. 1982. "*Mechanisms* for the Incorporation of Proteins in Membranes and Organelles." *Journal of Cell Biology* 92: 1–22.

Sattinger, Michael. 1993. "Assignment Models of the Distribution of Earnings." *Journal of Economic Literature* 31 (2): 831–81.

Savellos, Elias, and Yalcin, Umit., eds. 1995. *Supervenience: New Essays*. Cambridge: Cambridge University Press.

Schaffner, K. 1969. "The Watson–Crick Model and Reductionism." *British Journal for the Philosophy of Science* 20: 325–48.

———. 1976. "Approaches to Reduction." *Philosophy of Science* 34: 137–47.

———. 1993. *Discovery and Explanation in Biology and Medicine*. Chicago: University of Chicago Press.

Schmitt, Frederick, ed. 1994. *Socializing Epistemology*. Lanham, Md.: Rowman & Littlefield.

Schumpeter, Joseph. 1953. *History of Economic Analysis*. New York: Oxford University Press.

Searle, John. 1981. "Minds, Brains, and Programs." In *Mind Design*, ed. J. Haugland. Cambridge, Mass.: MIT Press.

Sent, Esther-Mirjam. 1997. *Resisting Sargent*. Cambridge: Cambridge University Press.

Sklar, Lawrence. 1993. *Physics and Chance*. Cambridge: Cambridge University Press.

Smart, J. C. C. 1989. *Our Place in the Universe*. Oxford: Basil Blackwell.

Smith, Stephen. 1994. *Labour Economics*. London: Routledge.

Snyder, S. 1985. "The Molecular Basis of Communication Between Cells." *Scientific American* 253: 132–52.

Sober, Elliott. 1984. *The Nature of Selection*. Cambridge: MIT Press.

Solomon, Miriam. 1994. "A More Social Epistemology." In *Socializing Epistemology,* ed. F. Schmitt. Lanham, Md.: Rowman & Littlefield.

Stich, S. 1985. *From Folk Psychology to Cognitive Science*. Cambridge: MIT Press.

———. 1990. *The Fragmentation of Reason*. Cambridge: MIT Press.

Tirole, Jean. 1982. "On the Possibility of Speculation under Rational Expectations." *Econometrica* 50: 1162–82.

Tonegawa, S. 1985. "The Molecules of the Immune System." *Scientific American* 253: 122–32.

van Fraassen, Bas. 1980. *The Scientific Image*. Oxford: Oxford University Press.

Walter, P., Ibrahimi, I., and Blobel, G. 1981. "Translocation of Proteins Across the Endoplasmic Reticulum." *Journal of Cell Biology* 91: 545–56.

Watkins, John. 1973. "Methodological Individualism: A Reply." In *Modes of Individualism and Collectivism*, ed. J. O'Neill. London: Heineman.

Weber, K., and Osborn, M. 1985. "The Molecules of the Cell Matrix." *Scientific American* 252: 100–110.

Weintraub, Roy. 1979. *Microfoundations: The Compatibility of Microeconomics and Macroeconomics.* Cambridge: Cambridge University Press.

Wilber, Charles. 1978. "The Methodological Basis of Institutional Economics: Pattern Model, Storytelling, and Holism." *Journal of Economic Issues* 12: 61–89.

Williamson, Oliver. 1975. *Markets and Hierarchies*. New York: Free Press.

———. 1985. *The Economic Institutions of Capitalism.* New York: Free Press.

Willis, R. 1986. "Wage Determinants: A Survey and Reinterpretation of Human Capital Earnings Functions." In *Handbook of Labour Economics*, ed. O. Ashenfelter and R. Layard. Amsterdam: North Holland.

Wimsatt, W. 1976. "Reductionism, Levels of Organization, and the Mind-Body Problem." In *Brain and Consciousness*, ed., G. Globus. New York: Plenum.

Yoshida, R. 1977. Reduction in the *Physical Sciences.* Halifax: Canadian Association for Publishing in Philosophy.

Index

About the Author

Harold Kincaid is professor of philosophy at the University of Alabama at Birmingham. He is the author of *Philosophical Foundations of the Social Sciences: Analyzing Controversies in Social Research* (Cambridge University Press).